AF531493

Mechanical Tools of Fabric Construction

Mechanical Tools of Fabric Construction

Sandeep Roy

RANDOM PUBLICATIONS
NEW DELHI (INDIA)

Mechanical Tools of Fabric Construction

ISBN 978-93-5111-869-5

Published in 2016 in India by

RANDOM PUBLICATIONS

4376-A/4B, Gali Murari Lal, Ansari Road
New Delhi-110 002
Phone : +9111-43580356, 011-23289044, 011-43142548
e-mail: sales@randompublications.com,
info@randompublications.com, randomexports@gmail.com

Type Setting by : Friends Media, Delhi-110089
Printed at : Sanat Printers

Preface

Constructing using textile materials require particular techniques and procedures to be undertaken to enable materials to be skilfully and safely measured, cut, shaped, joined and finished to make quality products. Advanced and complex techniques are required to craft special features of a high standard in a product and rely on the consistent application to achieve a desired effect. Special features, structural, style and/or decorative, include such things as set in sleeve, fly front, tailored collars and cuffs, welt pocket, embroidering, shirring.

Complex procedures include but are not limited to: joining materials with different properties, for example jacket shell and lining; changing the characteristics of the materials for example interfacing, interlining, boning, applied design; managing special fabrics, for example fine knits, sheers, satins; or designs cut on the bias.

Textiles are important for everyone. It is used for covering body, for warmth or coolness, personality enhancement and sometimes to display one's status in the society. From the wholesale textile manufacturer and merchant to the retailer and the end- user, the customer, everyone consumes textile. Not only those who are in this direct trade are related to this product but there are certain industries which are indirectly associated with textile. Automobile industry is a good example of this type of industry which uses textile in various forms. Others who use textile in one or the other form may include designers, interior decorators, craftspersons, advertisers using hoardings and banners, painters etc.

The term fabric can be defined as a planner structure produced by interlaced/interlooped yarns or fibers and felts made by interlocking fibers. It is a manufactured assembly of fibers and/or yarns that has substantial surface area in relation to its thickness and sufficient mechanical strength to give the assembly inherent cohesion. Basically, there are three methods by which fabrics are made.

This work is essentially useful for scholars researchers social activists academics government functionaries and the general reader alike.

– Author

Contents

1

Textile Fabrics

As far as the recognition of textine fibre is concerned it is considered to be very important as well as interesting subject. In modern world different kind of fibres have been discovered. Present day methods and processes adopted in the manufacture and finishing of fabrics, are making their identification more and more difficult. Moreover the consumer is interested in knowing the fibre content, which is eventually going to affect the durability, appearance, comfort and suitability for laundering and usage.

However, a knowledge of the characteristics of each kind of fibre helps a good deal, as every fibre has certain inherent qualities, which cannot be wholly reproduced in any imitation fibre. For example, wool has its own features and no other fibre has identical ones.

The look and even the touch of pure wool flannel may be copied in the cotton flannelette and an inexperienced purchaser may find it difficult to distinguish the wool from the cotton. Fabrics made with newer blends, synthetic fibres are also not easy to judge merely by the appearance or touch. To identify such fabrics, the help of other tests, even of chemical tests is needed.

FEATURES OF FIBRES

Natural Fibres

Man-made Fibres

These fibres do not take place in fibre form but have been turned into it by man, by breaking down from their original form and reassembled into various sort of structure, *e.g.,* rayons.

Cellulose-Fibres

These fibres have as their origin cellulose which is the chief matter of the plant cells. Cellulose is a complex compound made up of carbon, hydrogen and oxygen with the molecular formula of $(C_6H_{10}O_6)n$.

Fig. Cellulose Fibre

Cotton comprises about 91 per cent and hemp rame and flax contain similarly large amounts of cellulose. Cellulose is very sensitive to the action of mineral acids, and oxidising agents. However, it is very resistant to alkalies, including strong caustic alkalies at high temperature and pressure.

Cellulosic fibres are low in resiliency so the fabrics wrinkle easily. Due to the high absorbency of the fibre they are comfortable for summer wear.

Animal-Fibres

Wool and silk are obtained from animals and are, therefore, known as animal fibres. They have protein as one of the chief constituents and are made up of carbon, hydrogen and nitrogen. Wool contains sulphur besides the above elements. Both are destroyed by concentrated mineral acids but the specific action of the dilute acids is not very harmful on either. Thus, dilute acids are the basis of dyeing and finishing processes for the materials made from these fibres. Wool as well as silk burn with the special odour of burning proteins like burning of hair, dals or milk.

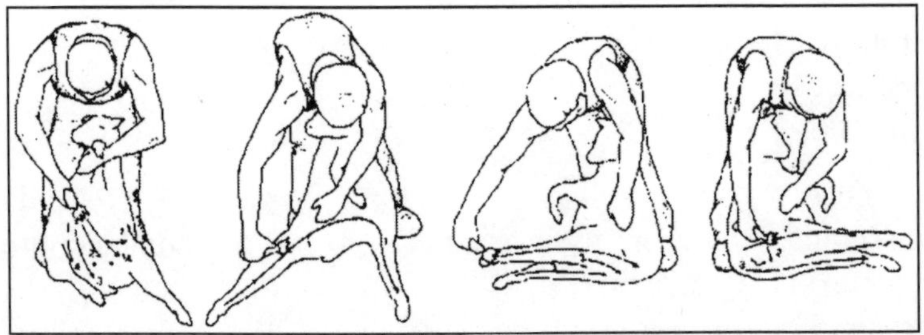

Fig. Animal-Fibres

Alkalies have a harmful effect on both. Animal fibres are very resilient and the wrinkles go out between wearings. However, they are bad conductors of heat and build up static electricity in cold and dry weather. Wool loses about 40 per cent of its strength when wet and silk, 15 per cent. Wool has a natural crimp which supports to increase its elasticity and strength. It has a springiness or resilience which no other natural fibre possesses. commonly speaking, woollen fabrics are soft to the touch and present a fuzzy appearance.

Worsteds are an exception, as their yarns are tightly twisted and firm. Wool, mixed with cotton, produces in the fabric a distinct hardness and heaviness about the material, which further tends to soil readily, fade and to wrinkle. Silk has a luxurious soft texture, and a deep lustre. The sheen is not loud or bright as in man-made fabrics.

This fibre is strong and retains most of its strength even when wet. It possesses elasticity and does not wear out soon. It is the lightest of all the fibres. Man-made fabrics may similiar to pure silk in appearance but are not so light. Silks weighted with metallic salts are still heavier and produce a sharp crease on the line of the fold. There are two kinds of man-made fibres: thermo-plastic and non-thermo-plastic.

Thermo-plastic Fibres

These are those that soften with heat and thus become liable or if the heat is sufficient high, will melt. These fibres are resistant to friction and wrinkling.

Fig. Thermo-plastic Fibres

They have low moisture content which makes them an easy to care fabric, however, they are not easy to dye. Due to this property they build up static electricity in cold and dry weather like animal fibres.

Non-Thermo-plastic Fibres

Which neither soften or melt with the application of heat but will scorch or burn if the temperature is very high. As a group they are soft, absorbent, pliable and comfortable to wear. They do not accumulate static electricity. They, except for mineral fibres, may be cared much as the natural fibres.

Fibres of Mineral Nature

These are inorganic and used for fireproof fabrics. Asbestos is practically the only natural mineral fibre. It is utilised in fabrics intended for make dress, fire-proof curtains and screens, and for man uses. It is also used for floor and table mats. Mineral fibres are really drawn threads from metals, *e.g.,* gold, silver in tissues and brocades, and even baser metals are utilised in fabrics like tinsel. Thus, their origin is inorganic matter.

They have an unusually bright sheen. Cotton is limp and has a dull surface because of to its fibre structure.

It lacks the lustre and the natural creaminess of linen. It is quite inflammable, and soils and crushes easily. It is heavier than comparable fabrics of other fibre content. It is brittle and is elastic when dry; but, when moist, its strength and elasticity increases. Mercerized cotton is stronger, smoother and lustrous, hence it soils less readily than cotton. Linen is smooth and cool to the touch as compared to cotton.

Fine linen has a lustre almost equal to that of silk. It has crispness and is strong enough for hard wear. Moisture soaks through and spreads over linen fabrics much more quickly than in the case of cotton fabrics.

Linen, when torn, has straight, glossy fibre-ends of unequal length. The fibre-ends of torn cotton curl and are lustreless. Linen yarn when not twisted, a more or less parallel arrangement of glossy, individual fibre.

Rayon is produced to replace pure silk and so has a similar appearance, but with a better luster generally; though sometimes dull rayons are also produced. The rayon fibre if subtly examined, will be found to be coarser and heavier than a silk fibre. Rayon realised stiffer to the touch and is less elastic. It breaks more easily than silk, thus showing less strength. It loses strength when wet and regains it when dry.

CLASSIFICATION OF TEXTILE FIBRES

Textile fibres are normally broken down into two main classes, natural and man-made fibres. All fibres which come from natural sources (animals, plants, etc.) and do not require fibre formation or reformation are classed as natural fibres. Natural fibres include the protein fibres such as wool and silk, the cellulose fibres such as cotton and linen, and the mineral fibre asbestos. Man-made fibres are fibres in which either the basic chemical units have been formed by chemical synthesis followed by fibre formation or the polymers from natural sources have been dissolved and regenerated after passage through a spinneret to form fibres.

Those fibres made by chemical synthesis are often called synthetic fibres, while fibres regenerated from natural polymer sources are called regenerated fibres or natural polymer fibres. In other words, all synthetic fibres and regenerated fibres are man-made fibres, since man is involved in the actual fibre formation process.

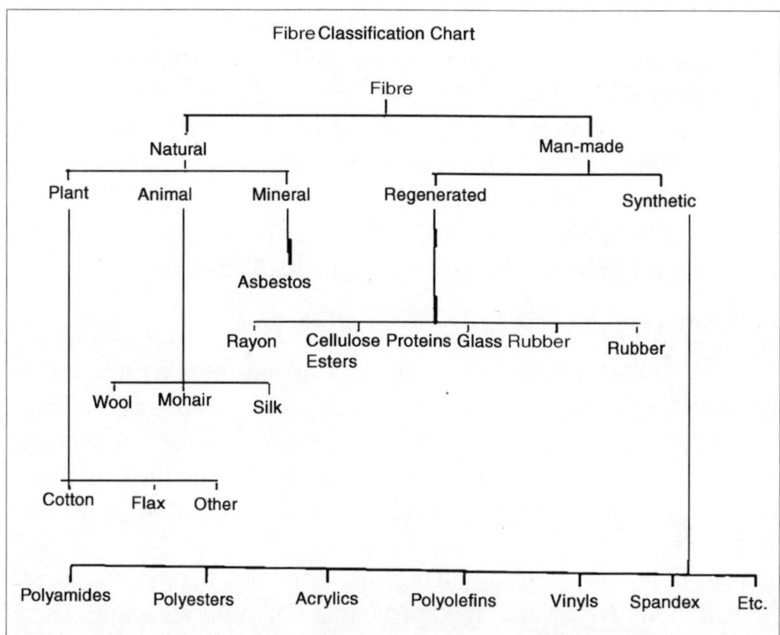

Fig. Classification of Natural and Man-made Fibres.

In contrast, fibres from natural sources are provided by nature in ready-made form. The synthetic man-made fibres include the polyamides (nylon), polyesters, acrylics, polyolefins, vinyls, and elastomeric fibres, while the regenerated fibres include rayon, the cellulose acetates, the regenerated proteins, glass and rubber fibres.

Another method of classifying fibres would be according to chemical structure without regard of the origin of the fibre and its starting materials. In this manner all fibres of similar chemical structure would be classed together. The natural man-made fibre classification given in figure does this to a certain extent. In this way, all fibres having the basic cellulosic unit in their structures would be grouped together rather than separated into natural and man-made fibres. This book essentially presents the fibres in groups of similar basic chemical structure, with two exceptions. In one case the elastomeric fibres have been grouped together due to their exceptional physical property, high extensibility and recovery.

TEXTILE AND POLYMER ENGINEERING PERSPECTIVE

The textile and polymer engineer must consider a number of criteria essential for formation, fabrication, and assembly of fibres into textile substrates. Often the criteria used will be similar to those set forth above concerning end-use properties.

Ideally a textile fibre should have the following properties:

- A melting and/or decomposition point above 220°C.

- A tensile strength of 5 g/denier or greater.
- Elongation at break above 10 per cent with reversible elongation up to 5 per cent strain.
- A moisture absorptivity of 2 per cent-5 per cent moisture uptake.
- Combined moisture regain and air entrapment capability.
- High abrasion resistance.
- Resistance to attack, swelling, or solution in solvents, acids, and bases.
- Self-extinguishing when removed from a flame.

FIBRE FORMATION AND MORPHOLOGY

Fibre morphology refers to the form and structure of a fibre, including the molecular arrangement of individual molecules and groups of molecules within the fibre. Most fibres are organic materials derived from carbon combined with other atoms such as oxygen, nitrogen, and halogens. The basic building blocks that organic materials form as covalently-bonded organic compounds are called monomers.

Covalent bonds involve the sharing of electrons between adjacent atoms within the monomer, and the structure tons between adjacent atoms within the monomer, and the structure of the monomer is determined by the type, location, and nature of bonding of atoms within the monomer and by the nature of covalent bonding between atoms. Monomers react or condense to form long-chain molecules called polymers made up of a given number (n) of monomer units which are the basic building unit of fibres.

On formation into fibres and orientation by natural or mechanical means the polymeric molecules possess ordered crystalline and non-ordered amorphous areas, depending on the nature of the polymer and the relative packing of molecules within the fibre. For a monomer A the sequence of events to fibre formation and orientation would appear as shown in figure.

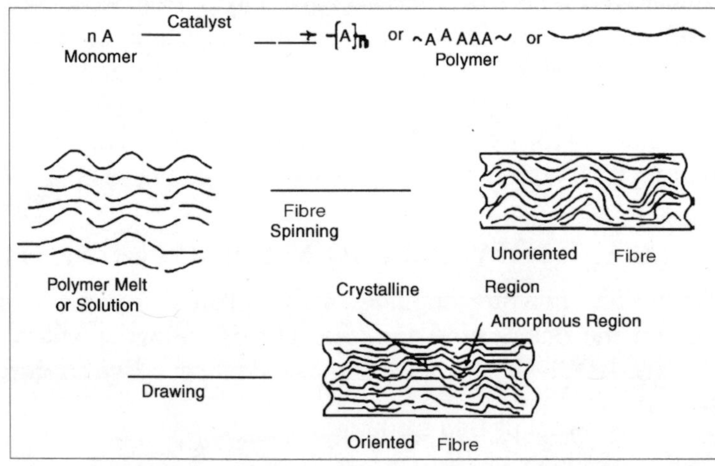

Fig. Polymerisation Sequence and Fibre Formation.

Polymers with repeating units of the same monomer (An) would be referred to as homopolymers. If a second unit B is introduced into the basic structure, structures which are copolymers are formed with structures as outlined in figure.

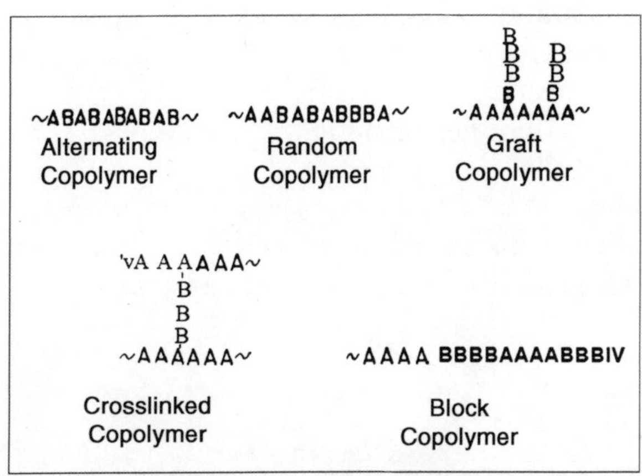

Fig. Copolymer Structures.

NATURAL FIBRES AND FIBRE SPINNING

Although natural fibres come in a morphological form determined by nature, regenerated and synthetic man-made fibres can be "tailor-made" depending on the shape and dimensions of the orifice (spinning jet) that the polymer is forced through to form the fibre. There are several methods used to spin a fibre from its polymer, including melt, dry, wet, emulsion, and suspension spinning. Melt spinning is the least complex of the methods. The polymer from which the fibre is made is melted and then forced through a spinneret and into air to cause solidification and fibre formation. Dry and wet spinning processes involve dissolving the fibre-forming polymer in an appropriate solvent, followed by passing a concentrated solution (20 per cent- 50 per cent polymer) through the spinneret and into dry air to evaporate the solvent in the case of dry spinning or into a coagulating bath to cause precipitation or regeneration of the polymer in fibre form in the case of wet spinning.

There is a net contraction of the spinning solution on loss of solvent. If a skin of polymer is formed on the fibre followed by diffusion of the remainder of the solvent from the core of the forming fibre, the cross-section of the fibre as it contracts may collapse to form an irregular popcorn like cross section. Emulsion spinning is used only for those fibre-forming polymers that are insoluble. Polymer is mixed with a surface-active agent (detergent) and possibly a solvent and then mixed at high speeds with water to form an emulsion of the polymer. The polymer is passed through the spinneret and into a coagulating bath to form the fibre.

In suspension spinning, the polymer is swollen and suspended in a swelling solvent. The swollen suspended polymer is forced through the spinneret into dry hot air to drive off solvent or into a wet non-solvent bath to cause the fibre to form through coagulation.

The spinning process can be divided into three steps:

1. Flow of spinning fluid within and through the spinneret under high stress and sheer.
2. Exit of fluid from the spinneret with relief of stress and an increase in volume (ballooning of flow).
3. Elongation of the fluid jet as it is subjected to tensile force as it cools and solidifies with orientation of molecular structure within the fibre.

Common cross sections of man-made fibres include:

- Round,
- Trilobal,
- Pentalobal,
- Dog-bone, and
- Crescent shapes.

When two polymers are used in fibre formation as in bicomponent or biconstituent fibres, the two components can be arranged in a matrix, side-by-side, or sheath-core configuration. Round cross sections are also found where skin formation has caused fibre contraction and puckering (as with rayons) has occurred or where the spinneret shape has provided a hollow fibre. Complex fibre cross-sectional shapes with special properties are also used.

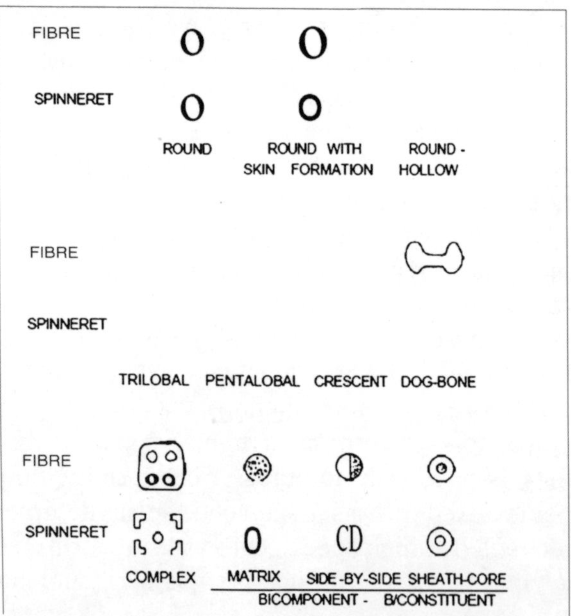

Fig. Fibre Cross Sections.

FIBRE DRAWING AND MORPHOLOGY

On drawing and orientation the man-made fibres become smaller in diameter and more crystalline, and imperfections in the fibre morphology are improved somewhat. Side-by-side bi component or bi constituent fibres on drawing become wavy and bulky.

In natural fibres the orientation of the molecules within the fibre is determined by the biological source during the growth and maturity process of the fibre.

The form and structure of polymer molecules with relation to each other within the fibre will depend on the relative alignment of the molecules in relationship to one another. Those areas where the polymer chains are closely aligned and packed close together are crystalline areas within the fibre, whereas those areas where there is essentially no molecular alignment are referred to as amorphous areas. Dyes and finishes can penetrate the amorphous portion of the fibre, but not the ordered crystalline portion.

SYNTHETIC POLYMERS FORMATION

Synthetic polymers used to form fibres are often classified on the basis of their mechanism of polymerisation—step growth (condensation) or chain growth (addition) polymerisation. Step growth polymerisation involves multifunctional monomers which undergo successive condensation with a second monomer or with itself to form a dimer, which in turn condenses with another dimer to form a tetramer, etc., usually with loss of a small molecule such as water.

Chain growth involves the instantaneous growth of a long molecular chain from unsaturated monomer units, followed by initiation of a second chain, etc. The two methods are outlined schematically:

$$\text{Step Growth:} \quad nA \rightarrow \frac{n}{2}AA \rightarrow AAAA \rightarrow$$

$$\text{Chain Growth:} \quad nA \rightarrow (A)n \rightarrow (A)n \qquad nA \rightarrow (A)n$$

The average number of monomer repeating units in a polymer chain (n) is often referred to also as the degree of polymerisation, DP. The DP must exceed an average 20 units in most cases to give a polymer of sufficient molecular size to have desirable fibre-forming properties. The overall breadth of distribution of molecular chain lengths in the polymer will affect the ultimate properties of the fibres, with wide polymer size distributions leading to an overall reduction of fibre properties. Although the polymers from natural fibres and regenerated natural fibres do not undergo polymerisation by the mechanisms found for synthetic fibres, most natural polymers have characteristic repeating units and high degrees of polymerisation and are related to step growth polymers. Basic polymeric structures for the major fibres are given in figure.

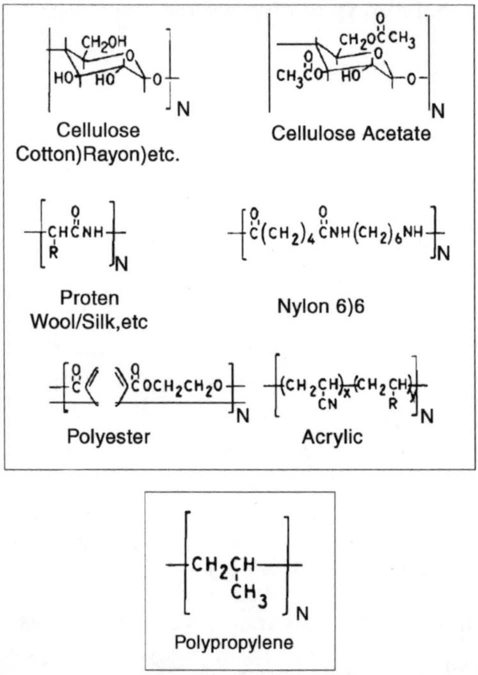

Cellulose
Cotton)Rayon)etc.

Cellulose Acetate

Proten
Wool/Silk,etc

Nylon 6)6

Polyester

Acrylic

Polypropylene

Fig. Basic Polymeric Structures for Major Fibres.

DEVELOPMENTS IN FIBRE MATERIALS - NATURAL FIBRES

Until early in the 20th century, the major fibres available for technical and industrial use were cotton and various coarser vegetable fibres such as flax, jùte and sisal. They were typically used to manufacture heavy canvas-type products, ropes and twines, and were characterised by relatively heavy weight, limited resistance to water and microbial/fungal attack as well as poor flame retardancy.

Some of the present day regional patterns of technical textiles manufacturing were established even then, for example Dundee, on the east coast of Scotland and located at the centre (then) of an important flax growing area as well as being a whaling port. Following the discovery that whale oil could be used to lubricate the spinning of the relatively coarse jute fibres then becoming available from the Indian subcontinent, jute fabrics were widely used for sacking, furniture and carpet manufacturing, roofing felts, linoleum flooring, twine and a host of other applications.

Although its jute industry was to decline dramatically from a peak at around 1900 owing to competition from other materials as well as from cheaper imports, Dundee and the surrounding industry subsequently become a nucleus for development of the UK polypropylene industry in the 1960s. The then newly available polymer proved not only to be an ideal technical substitute for the natural product but was also much more consistent in terms of its supply and price.

Traditional end-uses for sisal were similarly rapidly substituted throughout the established rope, twine and net making centres of Europe and America. Wool proved far less versatile and economic for most industrial applications although it is still valued for its insulating and flame retardency properties and finds use in several high temperature and protective clothing applications. Silk is an even more exotic fibre, rarely used in technical applications other than for highly specialised uses such as surgical suture thread. However, the traces of the early silk industry are still to be seen in the present day location of centres for technical filament weaving such as the Lyons area of France. The traditional silk industry has also contributed to the development of technical textiles in Asia, especially in Japan.

Textile Processes

The wide range of processes employed in the manufacture of technical textiles. Apart from the use of plaiting and knotting for the manufacture of ropes and nets, weaving was, for many years, the pre-eminent technology employed in the manufacture of 'industrial' textiles. In terms of the total weight of textiles produced, weaving still plays a leading role and developments such as three-dimensional and crimpless weaving have opened up many new product and end-use possibilities.

However, the historical progress of technical textiles has seen the advance of alternative textile forming technologies, most prominently the broad family of non-woven techniques but also warp and weft knitting, stitchbonding and modern braiding methods. The use of loose fibres with sophisticated cross-sectional profiles for insulation, protection and fibrefill applications is another important growth area. Fibres, yarns and textiles of all types also provide the starting point for a diverse and fast expanding range of composite reinforcement and forming technologies.

According to a major study of the world technical textiles industry and its markets projected to 2005, nonwovens are set to overtake weaving (in terms of the total weight of textiles produced) by around 2002/2003. In area terms, nonwovens already far exceed woven and other fabric forming methods because of their lower average weight per unit area. On the other hand, woven and other yarn-based fabrics will remain in the lead in value terms, at least for the foreseeable future. There is, therefore, something for every section of the textile industry in the future of technical textiles. Most product areas will see more rapid growth in value

than in volume as technical textiles become increasingly sophisticated and employ more specialised and higher value raw materials. On the other hand, the total value of yarns and fibres and of all technical textile products will grow slightly less fast than their volume because of a changing mix of materials and technologies, especially reflecting the growth of nonwovens.

FABRIC PERCEPTIONS IN TEXTILE

In recent years there has been a dynamic shift in attitudes toward textiles, and a new type of textile designer has challenged traditional forms of construction and content. As a result of a broadening vocabulary of materials, processes and techniques, textile designers are faced with more creative choice than ever before. New sensitivities and responses are also called for. On another level, attitudes and activities within the practice now mirror all facets of contemporary life; textile designers can be radical, funky, technical, conservative or serene; tradition has loosened. Many of the exciting changes that have happened in fabric in recent years have acted as a creative stimulus for other design disciplines.

This versatile medium now supplies the worlds of fashion and interiors with an increasing and sometimes challenging range of materials and products. The fashion designer now constantly looks for original, innovative materials to work with, making fabric a fundamental part of contemporary fashion statements.

For the interior designer or architect, new textiles offer everything from the complete co-ordinated chilled look to eclectic zany mixes. Whether they are looking for harmony, drama or functionality there is always one or more of these on offer, and quite often all three of them at the same time.

An increasingly diverse and sophisticated consumer market has witnessed innovations in fabric blends, the use of new technologies and the amazing march of synthetics. The effect of this combination has been the ability to explore new surface qualities of matt and shine as well as new combinations of weight and performance, and to reinvigorate old yarns and traditional fabric patterns. The look, the feel and the drape of fabric has entered a new era. No longer is the use of materials restricted to a perennial function: sportswear enters street wear, safety and protective clothing can influence couture. These dynamic fusions in their own way stimulate further opportunities - there are new surfaces to print on, and fabric development starts to play with surface qualities as variable as the difference between paper and rubber, between moss and cellophane.

New Fabrics, New Aesthetics

The fusion of technology and tradition has revealed itself in a diverse assortment of new fabrics and materials, impeccably mirroring changes in lifestyle and the way we embrace modernity and progress in the twenty-first century. Increasingly the gap between the producer of textiles and the consumer is shrinking, enforcing the demand for innovation. The merger of tradition and technology applies to almost every aspect of the textiles world; from designers to craftspeople to artists and industry.

Lifestyle culture has also been responsible for blurring the relationship between these disciplines and creating new intermediate markets and opportunities. Similar changes in the world of fashion create even more hybrid designers and virtuoso applications of fabric. Consequently, during the last decade of the twentieth century, a new breed of designers came to the forefront of textiles, and their work sits as comfortably in a gallery as it does on the shelf in Barneys, Liberty, Bergdorf Goodman or Harvey Nichols. The 'new' textile professional has foresight and employs intuition, working in a space where there are no limitations or restrictions.

Advances in textile design have had a remarkable effect on fashion. In addition to exploring issues of silhouette, the fashion designer envisages material innovation as an important motivating factor in his or her creative process. On occasion the material itself may well inspire silhouette. The creative collaboration between the fashion designer Issey Miyake and Makiko Minagawa, the textile director of Miyake Design Studio, illustrates this. Today's textile designer is resourceful and multi-talented, and can be found either in a laboratory breaking new ground through developing hi-tech materials, or leading the way by broadening the vocabulary of advanced decorative effects.

This is often achieved by means of investigating processes and treatments such as moulding, embossing, heat-bonding and sculpting as practiced by Sophie Roet and Nigel Atkinson. Material innovation in textiles involves surface exploration which can now result in novel, sometimes curious and extraordinary fabrics. These are attained by way of combining various methods of production and finishing. Outcomes can be exciting but poised carefully between the worlds of commercialism and discovery.

More unconventional approaches to making and designing are often stimulated by developments in yarn and fibre technology. These constantly enhance the creativity of the textile designer, as do innovations in textile processes which direct us away from traditional notions about the subject. Yarn innovation has led to avant-garde mixtures of fibres, including the silk and aluminium 'Inox' yarn and the wool and stainless steel blend from 'Bekintex'. These blends not only boost the properties of natural fibres but also usually offer additional qualities such as reducing static. Besides possessing performance-enhancing benefits, yarns are also developed for aesthetic reasons in order to achieve specific visual or textural effects. Synthetic fibres are often treated to mimic the intrinsic qualities of natural fibres or to create unusual, quirky finishes such as crunchy, glassy, clear nylon or novelty rubberized wool that feels like paper and looks like bark.

The recent technical and aesthetic progress in textiles deserves wider recognition. Today, the subject includes remarkable fabrics, distinctive furnishings, fascinating artworks and spectacular fashion. We are witnessing an era that is historically unmatched in terms of invention and experimentation in the field of textiles. An increasing number of designers are revealing the scope of the subject and the versatility of textiles as a creative medium. Through its major trade fairs, scattered across every continent, textiles now manages to disseminate these innovations rapidly across the globe.

The Evolution of Textile Practice

Conventionally, textile designers will have a strong grounding in one or perhaps two of the traditional specialisms of woven, printed, knitted and embroidered textiles, and these are still considered fundamental skills in textile practice. However, now is perhaps the right time to question and dispute these traditional categorizations of fabric. After all, current material experimentation within the discipline takes the subject beyond the realm of the yarn and the cloth. Technological advances, sophisticated consumerism, changes in society and lifestyles, diverse innovative thinking and media and communication touch not only textiles practice and philosophy but also the wider sphere of design disciplines in general.

Textiles is closely implicated in disciplines as diverse as fashion, visual communications, product design and interior design. Attempting to maintain the separate quality of traditional textile practices has proved difficult: as the subject has evolved, there has often been contention over the value and purpose of maintaining the purity of techniques and approaches; calls for change, and calls against it, immediately face an onslaught of criticisms. Textiles is now obviously something greater than the craft disciplines it has grown from.

Given this, these elementary skills still remain the foundational experience of textile practice, and they provide the textile professional with an initial framework. However, what follows the acquisition of those first skills is rather more complicated to describe or constrain. Simply describing textiles as printed, constructed or embroidered is not adequate. Crossovers occur continually, and they are not only within the disciplines mentioned here: textile designers dabble in history, culture, industry and science. The language of textiles has thus expanded through the absorption of technical knowledge or artistic ambition drawn from other subjects. With inventiveness and vision, new dimensions have been applied to time-honoured skills and materials and a case can now be made for the redefinition of textile practice.

Isabel Dodd is one of the new generation of designers whose alternative approach to textile design securely positions her in the future. Working with the most modern of synthetic materials, her printing process causes the surface of stretch jersey and velour to crinkle and pucker, producing marks similar to those formed through embroidery. We could loosely categorize her as a printer, simply because she uses a print screen. However, she is actually a typical example of individual textile designers who transcend their original formal training in order to interpret and communicate new creative ideas.

We may think that jumping from one specialist practice to another is daunting, yet many contemporary textile designers use whichever methodology is most appropriate for them to realize their conceptual ideas. In many ways this method of working is far from being a hindrance - in fact, a greater knowledge of process, by and large, enhances creative aptitude. Dodd's fabrics have been described as unusual, curious, avant-garde and sculptural, but whichever way they are explained, they are not what we traditionally associate with the conventional tools she uses.

The Influence of Skill

Fusions of traditional skills and new technology question whether textiles practice sits more comfortably in the scientific and industrial arena or whether it is essentially hands-on and skills-based. The introduction of new technology raises the question as to whether the practice could be branded as either purely scientific or, more traditionally, as an expression of an aptitude with material. The new concerns within textile practice have encouraged a fresh approach to a traditional discipline; one that is now in transition and appeals for re-evaluation as amply demonstrated by the recent 'A Twist in the Yarn' conference organized by Alan Holmes at Manchester Metropolitan University.

There is currently much research taking place in the textile world, exploring technologies related to textile production, and design issues involving the relationship between surface and structure. There are also a number of textile professionals involved in material development and research who would challenge any predetermined views of what textiles constitutes. Sarah Taylor, Janet Stoyel and Janet Emmanuel are among those who are revolutionizing our perceptions of fabric.

Sarah Taylor has become somewhat of a pioneer in the field of weaving, having worked in collaboration with the Japanese company Mitsubishi Rayon. Her research centres on developing light-emitting fabrics through the use of fibre optics.

The close supportive relationship she has with industry is vital in the development of her technologically advanced woven fabrics. Despite the craft origins of Taylor's work, perhaps industry on more than one level has historical sympathies with her practice - after all, the implicit data at the heart of weaving is easily convertible for mass production in industry:

There is a fluidity in the practice, design and art of woven textiles that enables textiles to fit easily with contemporary technology. A textile maker or designer who works at a small craft-shop level producing one-off pieces can, from the same conceptual base and using the same equipment, produce samples that industry can convert without any fuss for factory production.

Fibre optics have been used in protective clothing for military use, but Taylor has been investigating their possible applications within interiors, perhaps as multifunctional mood-enhancing design pieces. Through controlling the colour of the light released from the fibres, Taylor proposes that it would be possible to agitate or calm the environment. Cheryl Adnitt is also developing methods of incorporating light within her fabrics.

Her approach differs from Taylor's in that she uses either translucent monofilament which glows when exposed to ultraviolet light, or weaves conductive wire and LEDs into her fabrics, which illuminate when connected to a power source. The latter part of the twentieth century witnessed a definite move in the direction of a rather more scientific, technological and laboratory-based approach to textiles. This combination of chemistry, physics, textiles and design has conjured up numerous ground-breaking fabrics.

Janet Stoyel's work fits into this new category of practice, and she would no doubt agree that industrial liaison is advantageous and sometimes essential when working on developing hi-tech processes through innovative design. Stoyel is recognized as leading the way in the use of lasers for producing textile designs. Conventionally, ultrasonic technology has been employed for various purposes in the medical and industrial worlds. On the whole, within the textile industry, it is a relatively new discovery, first appearing on the scene in the early 1970s. For Stoyel to practice her design skills, it was crucial that she should be able to demonstrate proficiencies in areas other than textiles. There was a significant learning curve before she was able to operate the sophisticated machinery she needed in order to create the sculptural pieces she envisaged. Stoyel's range of skills can be compared to that of other textile designers such as Dodd and Taylor, already mentioned.

Much of Stoyel's material explorations conform to her unorthodox rules in that she blends synthetic metal and polymer textiles with such natural materials as silk and paper. The UK Ministry of Defence assisted Stoyel in her efforts to unify textile design and technology. Through utilizing their expertise she was able to establish her process of cutting using lasers and welding with ultrasound in order to create three-dimensional constructions.

Following in Stoyel's footsteps is Janet Emmanuel; although they share common ground in the equipment they use, their reasoning for making and the culmination of their efforts is quite different. Emmanuel's initial interest in technology can be narrowed down to a period at the end of the twentieth century when creative professionals from all design disciplines began to exploit new methodologies and materials.

Her preliminary studies within the subject focused on origami techniques, initially in silk organza and then in synthetics: these evolved into pleating ideas which were explored through traditional and industrial means, the goal being for the forms to become permanent. Some may feel her personal inspiration is somewhat obscure, but the driving force behind her research was to explore sound. In contrast to twentieth-century artists such as Kandinsky and Yves Klein who were fascinated with the visualization of sound (interpreting the relationship between colour, music and mark), Emmanuel opted to translate and examine what she calls silent sound, manifesting this through material manipulation.

The Influence of Science and Engineering

It is very difficult to escape the fact that science and engineering play a huge part in textile industries; much of the textile 'engineering' that occurs is usually intended for the manufacturing sector, but can also greatly influence design. An observation of history reveals that there have been major technical advances which in time have proven to be revolutionary for designers. Among these have been man-made regenerated fibres, synthetics and more recently smart or intelligent fibres.

One reason behind the original development of materials that mimicked natural qualities was to provide the consumer with an alternative to natural fibre at a fraction of the cost. The purpose for developing these materials, which would act as a substitute for the authentic article, was at some point superseded by the idea that we developed fabric just because we could. The scientists and engineers were 'able to' enhance, improve on and surpass the previous advances, so they did. Issues like fabric performance were stimulated by specialist sectors such as the sportswear industry; coupled with increasingly sophisticated consumer requirements, conditions were perfect for developments to continue.

The first 'engineered' fibres were generated from organic material and were developed to mimic the qualities of natural yarns, enabling the production of artificial silk and wool fibres. Developed in the second half of the nineteenth century, these cellulose-based fibres, constituted from wood pulp, were chemically altered so as to compose yarn. Following this development many years later, under the influence of environmental pressures, a 'green' recyclable artificial fibre was launched, again based on wood pulp. Apart from the introduction of microfibres (very thin, fine versions of synthetic fibres), Tencel is the first new man-made fibre to have been introduced since Polyolefin in the 1960s. It is renowned in the industry for being the 'chameleon' of fibres as it is able to completely imitate the qualities of other fibres, at the same time enhancing their strength.

Synthetics on the other hand were developed in science labs from such substances as coal, oil, other minerals, glass and carbon, after the discovery of regenerated fibres. There are two prominent established players in the world of synthetics, which have now become household names, Du Pont and ICI. Much of the initial research underlying the development of synthetic materials took place in the 1930s, while manufacturing became firmly established in the 1950s. Many synthetic fibres such as nylon, acrylic and DuPont's famous brand Lycra are household names. Polyester is now combined with its many natural counterparts, and is the most commonly used synthetic fibre, recyclable from clear plastic bottles. Synthetics have become infinitely sophisticated since their introduction microfibres and elastanes have been developed which possess virtues of highperformance and handle that natural fibres would find difficult to match.

In comparison to the innovations already mentioned, smart fabrics are in a league of their own. The sportswear industries drive the textile researchers, designers and technologists to create increasingly sophisticated materials to meet their demands. On first impression the needs and desires of the end user in this sphere could quite easily be confused with the stuff of a science-fiction novel. Materials that react to the body's needs are of the utmost importance in the sports world.

More often than not, these materials are developed to enhance performance as in the case of the Sydney 2000 Olympic gold medal winner, Ian 'Thorpedo' Thorpe, who hit international headlines when he wore the Adidas hydro-dynamically styled 'Fastskin' swimsuit. This concept was actually originated by the Britain-based Speedo swimwear company to reduce drag in the water. They state that the conceptual basis for their material was sharkskin, and that 'Fastskin' was designed to imitate the way in which a shark's scales enable it to glide smoothly through the water. The theory of copying naturally occurring phenomena like this is not new and, as engineering skills increase, such copying is likely to happen more often.

An increasing trend is that fabrics and yarns developed for a specific application or industry are used by other product sectors. For example, fabric technologies originally intended for industrial use or medicine can now form a central part of sport or fashion. Both of these worlds are colossal industries in their own rights and it is difficult to ignore the volume of the textile components used within each of them. In order to indicate the scale of the industries we are examining, a recent study by the American Georgia Institute of Technology, revealed that sport alone is a bigger industry than a combination of the television, radio, motion picture and educational service industries. Dr Isa Hoffman, Project Manager at Interstoff, Frankfurt, Germany, sums up current opinion on smart materials:

It is the fabric that makes the trend. Garment physiology and environmental behaviour, comfort and lifestyle have become the components of modern fabric creations and turned them into high-fashion, high-tech products ... Words such as 'intelligent' or 'electrotextiles', 'anti-bacterial' or 'biomimetic' could well come from a computer dictionary. Today's fabrics are 'programmable'. In beginning to reappraise the current condition of the textile world, we note that the 'future model' of the material supplier may also be transformed. As the definitions within textile practice begin overflowing into the worlds of science, engineering and technology, the fashion designer, product designer or architect that seeks innovative, original and new materials will inevitably search for a source other than that of the conventional fabric fair.

One such 'future model' of the fabric and materials supplier is George M. Beylerian, founder of Material ConneXion. He offers the facility to track down the most advanced materials and their suppliers from all over the world through his resource exchange. Additionally, he provides the expertise to assist with market research, design sourcing and product development; in doing so, he connects the design community with its technological counterpart.

Within this model, Beylerian is keen to work with material craftspersons, believing their role to be that of creative philosophers, and using their various skills and methodologies to manipulate and control new materials. It is envisaged that the exploration and assessment of the intrinsic properties of these materials will reveal the scope and manner of their application. This knowledge would lead to further material evolution. The role of the material supplier, as Beylerian himself puts it, may well be as 'an intelligence bank for the future'.

Re-evaluation

Definitions within contemporary textile practice are now open to debate, and traditional boundaries between sub-disciplines have become blurred. No longer does a weaver just sit at a loom, or an embroiderer at a sewing machine. Today, such persons, while engaged in their creative process, are to be found in front of computer screens, in science labs or engineering departments, or working alongside specialists from other disciplines. The creative processes now found in textiles defy the historical divisions and definitions applied to it. Currently, many practitioners who consider themselves textile specialists are occupied with roles and tasks not conventionally linked to the textile world. Textiles is full of contradictions which actually contribute to the innate nature of the subject, and simultaneously slants towards craft and towards technology.

Complications and confusions can also arise in its literature: we find printed textiles and felt-making featured in embroidery journals, craft periodicals don't have a problem with presenting textile art, and pieces on embroidery are included in felt-making publications.

Since much textiles practice fuses invention, design, craft and technical skill, each discipline within the textile arena seems to fit snugly into another. This fusion is not some abstract theoretical hope - it is a self-evident reality, populated by real textile people. Among the many designers that fit into this rather blurred new categorization of '1/1/textile professional' is Kei Ito. She investigates materials, amalgamating organic structures with ultramodern fabrics, manipulating them to form clever, complicated outcomes. Ambiguities arise when an attempt is made to categorize the work, which sits easily between textiles, craft and fashion art. With the new approach to textile making come new approaches to evaluating the practice and the medium. Freddie Robbins attempts to intellectualize the physical realm of knitting and compares it to current technology, making observations about our perceptions

There are many similarities between knitting and the internet. They are both physically solitary processes.

Both are making sense of, translating and communicating information. Computers and knitting patterns use grids, with coloured squares to relay visual information. With both mediums it is easy to 'undo' or 'unravel' your work. However, knitting is an 'everyday skill' of the future which crosses social, cultural and geographic boundaries. The reappraisal of textiles has repercussions in education and in the definition of the textile industries in general. Over the last few decades of the twentieth century textiles gradually reformulated itself, perhaps not intentionally, but nevertheless it is now not what it used to be. This evolution is attributable to a new generation of inquisitive designers and the ready availability of exciting, versatile and innovative raw materials.

All this makes the terms print, weave, knit and embroidery suddenly quite archaic, and one could almost say obsolete, as the divisions between them are eradicated.

We are truly at a point in time where re-evaluation is needed; this is even reflected in the nature of the student experience within the subject, many no longer desiring pure and simple textile design. There is a craving for something altogether more definitive of the twenty-first century, textiles of the future. As a result of the perpetual changes in textiles technology and culture the impetus to redefine textiles will undoubtedly grow.

TESTS FOR FIBRES IDENTIFICATION

The following tests can be carried out without the help of a laboratory or costly chemicals by a student of Textiles or by a housewife, interested enough to find out for herself the genuine-ness of the fibres of the material she wishes to purchase:

Labels

If the label gives the fibre content and tells how to take care of the fabric for satisfactory wear the consumer is most well-equipped to recognise the textile fabric she is purchasing. Unluckily there is no standardisation in labelling in India and if at all there is any label the information is incom-plete.

Inspection of Visual Nature

The first step in identification of a fibre is always inspection of a fabric for appearance and realise of the hand. The ability to the fibre by mere touch needs long experience and keen perception. It covers visual and textile senses.

Grasp the edge cloth the thumb and the index finger with the thumb on the top. Rub the thumb and the fore-finger across the cloth, lengthwise and then in a circle. Feel for pliability, elasticity or "give", warmth, softness, smoothness, body and hand. The vegetable fibres are generally cooler to touch than animal fibres or synthetics, a hundred per cent cotton fibre without a finish is lifeless and without a hand. Linen is cool, and pliable silk is very soft, lustrous, smooth and slippery and wool is warm and resilient. Rayon fabric is cooler to touch than acetate. However, it is quite difficult to identify each fibre especially the synthetics by mere touch because the kind of yarn used, finish and construction can often change the "feel" of the fabric.

Sight's Sense

This should be combined with the sense of touch. Observe luster or sheen, fuzziness, flatness and coarseness of fabric.

SOME GUIDELINES FOR THE BURNING TEST

Ignite a Piece of Fabric and Note

- How readily it burns.
- Amount and shape of ash.
- Odour.
- Kind of flame.

Protein fibres silk wool and hamburn briefly and charblack crushable ash Burning hair self extinguishing. Ransic flax cottonburn readily and charFirm soft grey ash, smooth edge burning paperhot self-extinguishing. Orlon, acetate acrilanmelt and burnforms hard plastic black beadchemical odour acetate has a characteristic acetic odour. Dacron, nylon, rayon, dynelmelts and burnsformation of beadschemical odour.

Glass and asbestos—Do not burn at all.

Particular Identification

Physical tests

Have a yarn from the fabric, unravel it and note the length of the fibres. The length of the fibres gives an indication to the group it belongs to, like all natural fibres barring silk are staple fibres. If this test is combined with the feel and the appearance of the fabric it will support the consumer to be able to identify it.

Tearing Test (for cotton and linen)

For this test, a fairly large piece of material is required. It will be observed that:

- To tear a piece of linen, much more strength is needed than to tear a piece of cotton.
- When linen is torn, the sound produced is more actually sharp than when tearing cotton. The threads along the tom edge of linen are straight and look pointed, whilst resemble threads of torn cotton are curly, fluffy and brush-like.

Microscopic Test

A microscope study of the fibres is almost essential for the student of Textiles, as this is one of the most faithful ways of arriving at the correct identification of fibres. In cases of fabrics woven out of a mixture of fibres. It is very tough to get definite or clear results from any of the above mentioned tests.

The same problem arises in the case of cotton and linen, particularly when these fabrics are heavily sized or when cotton is mercerized. But a microscopic examination will not fail to provide a correct identification of the fibre or fibres utilised in a fabric.

Moisture Test (for cotton and linen)

In comparison to cotton linen absorbs moisture much more readily. Take a piece each of linen and cotton material and place a moistened finger under each piece. The linen piece will take up moisture much more readily than the cotton piece.

Burning Test

It is a simple but a very reliable test. It can be conducted along with the visual inspection while buging fabrics to be sure of the genuineness of the fabrics. It may not be able to identify a fibre in special but it can connote the group such as cotton and linen will have the same results when burnt.

Breaking test (for cotton and linen)

Unravel some threads from the selvedge and the weft of the material. Then single threads are held between the fingers and pulled, they break. This may be repeated once or twice.

A linen thread, being stronger, takes more strength to break than a cotton thread of the same thickness. Broken ends of the cotton thread are fluffy and brush like, and the filaments or fibres which make the thread are short. Regarding linen, the ends are pointed and sharp, and the filaments or fibres, are not so short as those of cotton.

Tear out a fine fibre from the material to be tested and mount it on an eye glass with either water or a mixture off lycerol and water in the proportion of 1: 10. Then fix the glass on the microscope and examine the fibre under high and low power and make a note of what you observe.

Ramie

This fibre resembles linen except that it shows much broader and more irregular markings.

Wool

In a cross section of the woollen fibre, two distinct layers will be noticed. The outer layer comprises overlapping scales and the inner is a continuous rod containing pure fluid. A longitudinal section of the fibre will show only the overlapping outer scales.

Cotton

It seems as a hollow tube with a thin cell wall flattened and twisted. Under high-power magnification, the structure appears to be complicated. The fibre exhibited a central canal surrounded by a wall made up of many concentric layers of material.

Silk

Silk is either "Cultivated' or is 'Wild' (uncultivated) Each will present a different picture under the microscope. Cultivated silk just unreeled from the cocoons is called 'raw silk'. After being boiled to remove some of its gum it is knownas 'pure white silk. The 'raw silk' under a microscope dissplays double rod-like filaments, covered with lumps of gum which give the fibres striated or cracked appearance. The 'pure white silk' is cylindrical or rod-like in appearance without a central canal, and with a smooth surface reflecting light. Wild silk differs from cultivated silk in look under the microscope. It is broad and flat, uneven in width and shows characteristic markings which run obliquely across the fibre.

Linean

This fibre dissplays a thick cell wall with a very narrow, central canal, which is visible only under high-power magnification. The fibre is characterized by peculiar markings known as 'nodes', which resemble the pointed ridges in a bamboo.

Synthetic Fibre

The viscose rayon exhibits its fibre to be corrugated and flattened with a little lustre. The cupram-moniums show round filaments of a bright lustre.

Nylon

Fibres are very fine and round, smooth and translucent.

Dacron

Fibres are straight, smooth and perfectly round. It as a characteristic speckled appearance.

Orlon

Flat, smooth and semi-dull.

Chemical Tests Solubility

These tests are the confirmatory tests along with the microscope test. They are of great significance in fibre identification, location of the fibres in mixture and also the quality of various fibres in blends, by dissolving out each component of a special fibre.

Cotton and Linen

- Cuprammonium Practical dissolution blue colour etc.
- Iodine. Silk and Wool 5 per cent Caustic Soda entirely dissolves wool Picric acid Permanent yellow colour etc.

Rayons Cuprammonium hydroxidewill completely dissolved ActeatedAcetone.

Nylon 90 per cent Phenol.

Airlian Con. Nitric acid.

Use of Chemical Test to Recognise Fibre

- Acrilan is insoluble in acetone formic and 77 per cent sulphuric acid and sodium hypochlorite. If will dissolve in dimethyl formamide.
- Acetate. If a sample of acetate is immersed in acetone, the acetate fibres are entirely dissolved, other fibres with the exception of Arnel and Dynel are not affected.
- Orion is not affected by common solvents like glacial acetic acid, chloroform, acetone and 88 per cent formic acid.
- Arnel. When the specimen is dissolved at room temperature for one hour in a mixture of 90 per cent methylene chloride and 10 per cent ethyl alcohol, Arnel and Acetate will be dissolved. Other fibres will not be affected. To make difference between Arnel and Acetate the samples are immersed in benzyl alcohol at 50°C for one hour. Arnel is unaffected and acetate dissolves.
- Creslan is not affected by glacial acetic acid, chloroform, acetone or 88 per cent formic acid. It can be distancted from Orlon by microscopic appearance.
- Nylon is not soluble in boiling sodium hydroxide, this distinguishes it from most other fibres. The only general solvents in which nylon is soluble are concentrated formic acid, phenol and cresol.
- Dacron. Like acetate and nylon, Dacron is soluble in metacresol, however unlike acetate, it is unsoluble in concentrated formic acid.
- Dynel is highly resistant to a great variety of inorganic acids, bases, salts, hydrocarbons and most organic solvents.

- Vicara is used mainly as a blending fibre, particularly with wool, cotton, rayon and nylon. It can be distincted from wool by caustic soda 10 per cent in which wool dissolves but vicara does not. Hydrochloric acid will distinguish vicara from cotton, rayon, acetate and nylon. The acid will dissolve other fibres except vicara.
- Saran. 100 per cent solution of acetone will not affect saran. Saran is soluble in ammonium hydroxide.
- Fibre glass dissolves in phosphoric acid and hydrofluoric acid. These are the only two acids which will affect fibre glass.
- Verel when placed in Pyridene and heated turns reddish brown but does not dissolve.
- Zefran has an affinity for a great variety of dyes. It may be distinguished by a Fastucal pink dye test. A bleached specimen dyed in 1 per cent solution of this dye in a neutral bath at room temperature for three minutes becomes pink, while other fibres remain white.
- Darvan will dissolve in dimethyl formamide at room temperature. It can be distincted from Acrilan by Calico identification stain 2 tints Darvan greyish pink.

2

Coloration of Technical Textiles

Technical textiles are used in a very wide variety of end-uses in which the functional performance requirements are paramount. Thus, technical textiles must possess the requisite physical and mechanical properties to maintain the structural integrity throughout all the manufacturing and fabrication processes and during the service life of the material. For some end-uses, therefore, coloration is not strictly necessary because the aesthetic properties of the technical textile, for example colour and pattern, and also lustre, texture, handle and drape, may not always be required to appeal to the visual and tactile senses of the customer.

Coloration of technical textiles by dyeing or printing is primarily intended for aesthetic reasons but also provides a ready means of identifying different qualities or fineness of the materials. For example, the fineness of a surgical suture and its visibility at the implant site are easily identified by colour. High visibility clothing and camouflage printing clearly provide the extreme ends of the coloration spectrum for technical textiles.

Coloration also introduces other functional properties distinct from the aesthetic appeal of colour. Colorants can hide fibre yellowing and aid fibre protection against weathering, both factors of importance where the physical properties of the technical textile must be maintained over a long service life under adverse end-use conditions. Heat absorption is also increased where black materials are exposed to sunlight, an important factor for baling materials for agriculture.

OBJECTIVES OF COLORATION

The objectives of coloration treatments are first to produce the desired colour in dyeing and colours in the coloured design image in printing on the technical textile. Second, coloration treatments are to ensure that the necessary colour fastness requirements for the end-use are achieved. Third, the whole operation should be carried out at the lowest cost commensurate with obtaining the desired technical performance. After dyeing, the levelness and uniformity of colour across the width and along the length of a technical textile fabric must be within the defined colour tolerances agreed between the dyer and the customer prior to coloration. Preservation of the original appearance and quality of the technical textile prior to coloration is also essential in order to ensure that the technical textile is of marketable quality.

TECHNICAL TEXTILES

Coloration involves the application of colorants to the technical textile and is a complex field because of the variety of fibres, filaments, yarns, fabrics and other materials requiring coloration and the diverse nature of the end-use and performance requirements. Coloration may be carried out by dyeing the materials to a uniform colour, or by printing to impart a design or motif to the technical textile. Fibres, yarns and fabrics may also be multicoloured by specialised dyeing techniques, for example space dyeing, or by weaving or knitting different coloured yarns.

The colorants used may be either water-soluble (or sparingly water-soluble) dyes, or alternatively water-insoluble pigments. Most dyes are applied to technical textiles in an aqueous medium in dyeing and in printing, although selected disperse dyes may also be dyed in supercritical fluid carbon dioxide (above the critical point under very high pressure) for a few specialised end-uses, for example polyester sewing threads. By contrast, pigments are either physically entrapped within the filaments during synthetic fibre extrusion, for example by mass pigmentation, to give a spun-dyed fibre, or alternatively adhered to technical textiles in pigment printing through the use of an adhesive binder.

Dyeing is normally carried out on textile materials from which surface impurities, for example fibre lubricants, spin finishes, sizes, particulate dirt or natural colouring matters and so on, have been removed by appropriate pretreatments (*e.g.* by desizing and scouring), and to which a stable whiteness has been imparted by chemical bleaching. However, many synthetic fibres do not normally require chemical bleaching prior to coloration because the fibres may be whitened by incorporation of a fluorescent brightening agent during fibre manufacture.

Printing may be carried out mainly on technical fabrics that may be in their natural state, or chemically bleached, or whitened with a fluorescent brightening agent, or after tinting or dyeing. Conventional dyeing and particularly printing are most conveniently and economically carried out on fabrics, which also allow greater flexibility through the selection of colours late in the technical textile production sequence to meet the varying market requirements.

Dyes

The dyes used commercially to dye technical textiles may be selected from the very wide range of synthetic organic colorants, based upon aromatic compounds derived from petroleum. Dyes are conjugated organic structures containing an alternating system of single and double bonds within the molecule which impart the ability to absorb certain wavelengths of visible light, so that the remaining light scattered by the dyed technical textile is perceived as coloured.

The dye structure must contain a chromophore, a chemical group that confers upon a substance the potential to becoming coloured, for example nitro, nitroso, azo and carbonyl groups.

To become a useful dye, however, the molecule should contain other chemical groups such as amino, substituted amino, hydroxyl, sulphonic or carboxyl groups which are called auxochromes. These generally modify or intensify the colour, render the dye soluble in water and assist in conferring substantivity of the dye for the fibre. High substantivity aids a high degree of dye exhaustion on to the fibres during exhaust dyeing when dyes are progressively and preferentially adsorbed by the fibres from the dyebath to give coloured technical textile materials. High dye-fibre substantivity also generally confers high colour fastness during end-use, for example high colour fastness when exposed to washing and light.

Manufactured dyestuffs are formulated colorants typically containing only 30-40% active colorant. The rest of the dyestuff formulation consists of diluents to dilute the colour strength, electrolytes to improve the dye exhaustion on the fibre, and dispersing agents to facilitate dispersing of the dye in the dyebath. To avoid dusting dyestuffs into the air during weighing and dispensing, granular, grain or pearl forms of dyestuffs are used. Antidusting agents are added to powdered dyestuffs for the same reason. Water-soluble dyes can be produced as true liquid dyestuffs (*e.g.* basic dyes), but liquid dispersions of disperse and vat dyestuffs are also produced.

The latter may contain a viscosity modifier to minimise sedimentation/settling of the dye particles during storage and must be stirred prior to use. Liquid dyestuffs are preferred for continuous dyeing and for printing because of their convenience for use on automated weighing, metering and dispensing systems.

Commercial dyestuffs are subjected to stringent quality control procedures for hue and colour strength, average particle size and particle size distribution, and are dried to a uniform moisture content and packaged in drums or plastic containers suitable for transportation, storage and dispensing. Resealable packaging may be used to prevent the ingress of moisture from the atmosphere, and in many colour kitchens the ambient temperature and relative humidity are controlled in order to maintain reproducibility in weighing. Some dyestuffs are mixtures formulated for specific shades such as black, or contain dyes from different application classes suitable for dyeing specific fibre blends, for example 65/35 polyester/cellulosic fibre blends.

Pigments

Pigments are synthetic organic or inorganic compounds that are insoluble in water, although some are soluble in organic solvents. The pigment particles are ground down to a fine state of subdivision (0.5-2 |jm) and stabilised for use by the addition of dispersing agents and stabilisers. Both powder and liquid (*i.e.* dispersions of) pigments are used for the coloration of technical textiles, and the pigment finish must be compatible with the application conditions used in mass pigmentation of manufactured fibres or in pigment printing of technical textile fabrics.

Fluorescent Brightening Agents

These are organic compounds that absorb some of the ultraviolet radiation in sunlight or other sources of illumination and re-emit in the longer wavelength blue-violet region of the visible spectrum. Such compounds can be applied to technical textiles either by mass pigmentation (in manufactured fibres) or via machinery used for conventional dyeing of all types of materials, and add brightness to the whiteness obtained from chemical bleaching of the textile.

Range of Colorants Available

The Colour Index International is the primary source of information on colorants and lists the majority of commercial dyestuffs and pigments (both past and present) although the situation continually changes as the manufacture of some colorants ceases, to be replaced by new products. The wide variety of colorants available is based upon many chemical types that make up the major application groups of colorants.

The percentage distribution of each chemical class between major application ranges, based upon all the dyes listed where the chemical class is known, but also including products which are no longer used commercially. Azo colorants make up almost two-thirds of the organic colorants listed in the Colour Index International.

Anthraquinones (15%), triarylmethanes (3%) and phthalocyanines (2%) are also of major importance. Where the structure of the colorant is known, it is assigned a specific CI (Colour Index) chemical constitution number, although for many new commercial products the structure is undisclosed.

In the Colour Index International dyes are classified by application class, by colour and by number, for example CI Acid Blue 45. The commercial name of the dyestuff will, however, vary according to the dye maker, and dyestuffs, being formulated products, although nominally the same CI number, may vary in active colorant content and in the nature of the additives incorporated within the formulation.

Metamerism may also be observed on the dyed textile, that is a difference in colour is observed when two colours are viewed under different illumination conditions, even when using dyestuffs that nominally possess the same colour index number. The Colour Index International is available in book form and on CD-ROM.

The major classes of dyes and the fibres to which they are applied. The type of dyestuff used for dyeing technical textiles depends upon the fibre(s) present in the material, the required colour and the depth of colour, the ease of dyeing by the intended application method and the colour fastness performance required for the end-use. Some dyestuff classes will dye a number of different fibres, but in practice one fibre will predominate. For example, disperse dyes are mainly used to dye polyester fibres, although they can also be applied to nylon 6, nylon 6.6, acrylic, modacrylic, secondary cellulose acetate, cellulose triacetate and polypropylene fibres, but with limitations upon the depth of colour and the colour fastness that can be attained.

There are, however, many types of manufactured fibres, particularly synthetic fibres, which are currently virtually impossible to dye with conventional dyestuffs using normal dye application methods. This is because filaments which are designed for high strength end-uses are often composed of polymer repeat units that do not contain functional groups that could act as dyesites (*e.g.* polyethylene, polypropylene, polytetrafluoroethylene), and/or may be highly drawn to produce a highly oriented fibre with a high crystallinity.

Because dyes are considered to diffuse in monomolecular form into the fibre only through non-crystalline regions, or regions of low order (*i.e.* high disorder), it follows that highly oriented fibres such as meta-aramid, para-aramid, and high strength polyethylene fibres, are extremely difficult to dye. In some cases, specialised dyeing techniques utilising fibre plasticising agents ('carriers') to lower the glass transition temperature (T_g) of the fibre (*i.e.* the temperature at which increased polymer seg-mental motion opens up the fibre structure) together with selected dyes may be employed to speed up the diffusion of dyes into the fibres, but the colour depth may be restricted to pale-medium depths.

In addition, or alternatively, resort may be made to high temperature dyeing in pressurised dyeing machinery at temperatures in the range 130-140 °C for the same purpose.

However, for many high strength fibres, mass pigmentation offers a more satisfactory production route to coloured fibres, provided that the introduction of the pigments does not significantly impair the high strength properties. Some fibres such as carbon fibres and partially oxidised polyacrylonitrile fibres cannot be dyed, simply because they are already black as a result of the fibre manufacturing process. However, contrasting colours may be printed on such fibres using pigment printing methods, or by blending with other dyed fibres.

Microfibres are generally defined as fibres or filaments of linear density less than 1 dtex. Such silk-like fibres pose considerable difficulties for level dyeing of medium to heavy depths of colour. The high surface area per unit volume of microfibres increases the light scattering, necessitating the use of greater amounts of dye to achieve the same colour depth as on coarser fibres. The use of bright trilobal filaments, or the incorporation of titanium dioxide delustring agents within the microfibres increases the light scattering, making the difference even more noticeable.

DYE CLASSES AND PIGMENTS

MORDANT DYES

Chrome dyes are the only type of mordant dye of any commercial significance. These are used in exhaust dyeing to dye wool or occasionally polyamide fibres to deep dull colours of high colour fastness to wet treatments and light. The fibres are usually dyed by the afterchrome method in which a chrome dye (similar to an acid dye) is dyed on to the fibre and then the dyed fibre is given a treatment at pH 3.5 in a second bath containing sodium or potassium dichromate.

The absorption of dichromate ions leads to the formation of a 1:1 and/or 1:2 chromium metal complex dye inside the fibre, which can lead to some problems in batch-to-batch repro-ducibility of shade, particularly if the pH control in chroming is variable. Low chrome dyeing procedures can promote exhaustion of the chromium on to the fibres, decreasing environmental pollution from the waste water from dyeing. A major disadvantage of the afterchrome method is that the final colour is not developed until the chroming stage which can give difficulties in shade matching. Ammonia aftertreatment can improve the colour fastness to washing.

Acid Dyes

Acid dyes are anionic dyes characterised by possessing substantivity for protein fibres such as wool and silk, and also polyamide (*e.g.* nylon 6 and nylon 6.6) fibres, or any other fibres that contain basic groups. Acid dyes are normally applied from an acid or neutral dyebath.

As the size of the acid dye molecule generally increases from level dyeing (or equalising) acid dyes to fast acid, half-milling or perspiration-fast dyes to acid milling dyes and thence to supermilling acid dyes, the colour fastness to washing increases because of the increasing strength of the non-polar forces of attraction for the fibre. 1:1 Metal complex and 1:2 metal complex dyes also behave like acid dyes from the viewpoint of application to the fibre, and possess good colour fastness to washing and light. Level dyeing acid dyes yield bright colours, while the larger sized milling acid and supermilling acid dyes are progressively duller. 1:1 and 1:2 metal complex dyes also lack brightness, but can provide good colour fastness performance.

Acid dyes contain acidic groups, usually sulphonate groups, either as —SO_3Na or —SO_3H groups, although carboxyl groups (—COOH) can sometimes be incorporated. Wool, silk and polyamide fibres contain amino groups (—NH_2) which in an acid dyebath are protonated to yield basic dyesites (—NH_3^+). The acid dye anion D.SO3, where D is the dye molecule, is thus substantive to the fibre and is adsorbed, forming an ionic linkage, a salt link, with the fibre dyesite. Monosulphonated acid dyes can be adsorbed to a greater extent than di-, tri-, and tetrasulphonated acid dyes because the number of basic dyesites in the fibres is limited by the fibre structure. Thus, the colour build-up is generally greatest with monosulphonated dyes on wool. However, the dyestuff solubility in water increases with the degree of dye sulphonation.

Dyed wool fibre quality is improved by pretreatment before dyeing with Valsol LTA-N(TM), an auxiliary which extracts lipids from the cell membrane complex of wool, thereby speeding up the intercellular diffusion of dyes. In this Sirolan (TM) LTD (low temperature dyeing) process, wool may be dyed for the conventional time at a lower temperature, or for a shorter time at the boil, decreasing fibre degradation in the dyebath.

Levelling agents may be used to promote level dyeing of acid dyes on wool and other fibres. Anionic levelling agents enter the fibre first and interact with basic dyesites, restricting dye uptake. Amphoteric levelling agents, which contain both a positive and a negative charge, block the basic dyesites in the fibre but also complex with the acid dyes in the dyebath, slowing down the rate at which the dyes exhaust on to the fibres. As the dyebath temperature is increased, the anionic (or amphoteric) levelling agent desorbs from the fibres, allowing the dye anions to diffuse and fix within the fibre. In addition, as the dyebath temperature is raised, the dye/ampho-teric levelling agent complex breaks down, releasing dye anions for diffusion inside the fibre. The dyes are thus gradually adsorbed by the wool as the dyebath temperature is increased.

In the iso-ionic region (pH 4-5), the cystine (disulphide) crosslinks in the wool are reinforced by salt links formed between charged carboxyl and amino groups in opposing amino acid residues in adjacent protein chain molecules within the wool fibre. This temporarily strengthens the wool and the abrasion resistance of the dyed fabric is improved. Fibre yellowing is decreased and brighter colours may be dyed.

The build-up of acid dyes on silk and polyamide fibres is limited by the fewer number of fibre dyesites compared with wool. Acid dyes are attracted to the amine end groups (AEG) in polyamide fibres such as nylon 6 and nylon 6.6. Nylon 6 has a more open physicochemical structure and a lower glass transition temperature (T_g) compared with nylon 6.6. Thus, acid dyes diffuse more readily into nylon 6 but the colour fastness to washing of a similar dye on nylon 6.6 will generally be superior because of the more compact fibre structure. False twist texturing processes, which use contact heating of the yarns, open up the fibre structure and may modify some of the fibre cross-sections which were in contact with the heater, resulting in a slightly lower colour fastness to washing.

Chemical variations due to changes in the AEG of nylon fibres, can give rise to dyeability variations with acid dyes, a problem often referred to as barré or barri-ness. Physical variations caused by temperature and/or tension differences in nylon fibres can similarly lead to differences in the uptake of disperse or 1:2 disperse premetallised dyes. Physical variations can be minimised by dyeing at higher temperatures (*e.g.* up to 120 °C) with nylon 6.6, or prolonging the dyeing time. High temperature dyeing is used with the larger 1:2 metal complex acid dyes in order to achieve better fibre levelling and also fibre penetration, which leads to improved colour fastness to washing. Acid dyes can also be printed on to wool (which has been pretreated by chlorination), degummed silk, polyamide and other fibres using conventional print thickeners in a print-dry-steam fixation-wash off and dry process.

Colour fastness to washing with acid dyes on nylon fibres, especially nylon microfibres, is improved by after-treatments such as syntanning, that is, the adsorption of a sulphonated synthetic tanning agent that provides a physical barrier to desorption by blocking the fibre pore structure in the fibre surface regions and by providing electrostatic (ionic) repulsion to dye desorption. Combining an appropriate syntan with a fluorochemical treatment can provide, in addition, stainblock-ing properties. However, syntan treatments degrade during high temperature or steaming treatments and tend to yellow the dyed fabric, dulling the colour.

Basic Dyes

Basic dyes are cationic dyes characterised by their substantivity for standard acrylic, modacrylic, basic-dyeable polyester and basic-dyeable nylon fibres. Basic dyes can be applied to protein and other fibres, for example secondary cellulose acetate, but the light fastness of basic dyes on hydrophilic fibres is poor. The major outlets for basic dyes are acrylic and modacrylic fibres, on which basic dyes can impart bright colours with considerable brilliance and fluorescence. The ionic attraction between the basic dye and the sulphonic acid dyesites in acrylic fibres is strong, which yields high colour fastness to washing. The close-packed physicochemical nature of acrylic fibres and the strong dye-fibre bonding can result in poor migration and levelling properties during dye application, but impart very high colour fastness to light.

Acrylic fibres vary widely in their dyeability because of the different amounts of different co-monomers used with poly(acrylonitrile) that modify the fibre glass transition temperature. This may range from 70-95 °C according to the source of the acrylic fibre manufacturer. Cationic retarders are widely used on acrylic and modacrylic fibres to promote level dyeing and the basic dyes used should preferably be selected with the same compatibility value. The compatibility value may range from 5 (slow diffusing) down to 1 (rapid diffusing). A compatibility value of 3 has been recommended for package dyeing, but for printing, compatibility values are of much less importance. Print fixation is by the print-dry-steam-wash off and dry process, although wet transfer printing techniques are also possible using selected dyes.

The acidic dyesites in acrylic fibres become much more accessible over a narrow band of temperatures in the glass transition temperature range. Therefore, even when using cationic retarders, which initially adsorb on to the fibre blocking the dyesites, the dyebath temperature must be raised slowly through this temperature range to allow the cationic retarder to desorb slowly and allow the basic dye to be adsorbed uniformly. Alternatively, constant temperature dyeing methods may be used in exhaust dyeing. Continuous pad-steam-wash off and dry methods can also be used for dyeing acrylic fabrics.

The basic dye uptake is limited by the number of acidic dyesites in the fibre, but approximately 95% of all colours on acrylic fibres are dyed using basic dyes. However, where very pale colours are to be dyed it is common to use disperse dyes, which have superior migration and levelling properties, in order to attain a level dyeing, because the strong dye-fibre bonding renders this very difficult to achieve using basic dyes. The build-up of disperse dyes on acrylic fibres is limited and the colour fastness to washing and light are generally lower than for basic dyes. The colour fastness of basic dyes on acrylic fibres is superior to similar dyeings on modacrylic fibres. Some modacrylic fibres are prone to delustring during dyeing and may require relustring by boiling for 30min in a high concentration of electrolyte (*e.g.* 50-200 gl^1 sodium chloride) or by dry-heat or treatment in saturated steam.

Reactive Dyes

Reactive dyes, sometimes termed fibre-reactive dyes, are a very important class of dyes for dyeing cellulosic fibres and are also used to dye protein fibres such as wool or silk. Although relatively expensive they provide a wide colour gamut of bright colours with very good colour fastness to washing. This is because during dye fixation, usually conducted under alkaline conditions, strong covalent bonds are formed between the dye and the fibre. Reactive dyes may react by substitution (*e.g.* monochlorotriazinyl and dichlorotriazinyl dyes) or by addition (*e.g.* vinylsul-phone dyes).

Typical ranges of reactive dyes and their respective reactive groups. Recent developments have led to the introduction of homobifunctional reactive dyes (*e.g.* two monochlorotriazine groups) and heterobifunctional reactive dyes (*e.g.* using monochlorotriazine plus vinyl sulphone) in an attempt to increase the dye fixation on the fibre under alkaline conditions from 50-70% with one reactive group, to 80-95% with two reactive groups. Dye application methods on fabrics include exhaust dyeing, cold pad-batch-wash off or continuous pad-steam-wash off and dry methods. Fixation by dry heat, saturated steam or superheated steam may also be used according to the type of reactive dye employed.

Reactive dyes hydrolyse in contact with water and alkali, and during dyeing some hydrolysed reactive dye is adsorbed by the fibre because it behaves like a substantive direct dye, but with lower colour fastness to washing than the reactive dye that is covalently bonded. Thus, the dyeing stage must always be followed by an extended washing-off treatment to remove hydrolysed reactive dye. Provided that the residual hydrolysed reactive dye left within the cellulosic fibre after dyeing or printing is at a concentration $<0.1gm^2$, the best colour fastness to washing will be obtained. If the concentration of hydrolysed reactive dye in the final hot wash bath is $<0.003gl\text{-}^1$, staining of white grounds in the fabric will be prevented.

Generally reactive dyes are produced in granular form,but vinyl sulphone dyes are also available as liquids, which are more convenient for use in continuous dyeing and printing. Reactive dyes are sulphonated and highly water soluble and are exhausted on to the fibre using electrolyte (*e.g.* in the manner of direct dyes) and fixed using an appropriate alkali. The high water solubility creates an environmental problem in waste water treatment plants because generally only 0-30% of the hydrolysed reactive dye is removed by such treatment. Low salt reactive dyes have also been introduced to decrease dyeing costs and to avoid corrosion problems in concrete waste water pipework networks caused by high concentrations of sulphate anions.

Azoic Colouring Matters

Azoic colouring matters are formed inside the fibre (usually cellulosic fibres) generally by adsorption of an aromatic hydroxyl-containing compound, such as a naphthol or naphtholate (azoic coupling component) followed by coupling with a stabilised aromatic diazonium compound (the azoic diazo component also termed fast base or salt) to form a coloured insoluble azo compound. The application method has to be carried out with care and the final colour is only obtained after soaping off to remove any traces of the azoic dye on the fibre surface that would give inferior colour fastness to washing and rubbing (crocking). Although they are economical for the production of red and black, the colour range is now more limited as some diazo components based upon certain aromatic amines have been withdrawn because of their possible carcinogenic nature. Precise colour matching with such a complex two-bath procedure can also give problems in practice.

Direct Dyes

Direct dyes were the first class of synthetic dyes to dye cotton directly without the use of a mordant. Direct dyes are sulphonated bisazo, trisazo or polyazo dyes and are anionic dyes which are substantive to cellulosic fibres when applied from an aqueous bath containing an electrolyte. Direct dyes based on stilbene, copper-complex azo, oxazine, thiazole and phthalocyanine structures are also used. Less bright in colour than acid or basic dyes, the brightness diminishes with the molecular complexity of the dye. Phthalocyanine dyes are used for very bright blue and turquoise-blue colours of good colour fastness to light, and copper-complex azo dyes also exhibit good fastness to light, although these dyes yield relatively dull colours.

Electrolyte, in the form of sodium sulphate, is usually added into the dyebath to overcome the negative charge on the cellulosic fibre surface which otherwise would repel the approach of the direct dye anions. The sodium cations from the electrolyte neutralise the negative charge at the fibre surface, allowing the dye anions to be adsorbed and retained within the fibre.

Coulombic attraction, hydrogen bonding and non-polar van der Waals forces may operate depending upon the specific direct dye and the fibre structure. Direct dyes are normally linear and planar molecules allowing multipoint attachment to cellulose chain molecules, but the forces of attraction between dye and fibre are relatively weak when the dyed fibre is immersed in water. Thus, the colour fastness to washing is moderate to poor, but the colour fastness to light may vary from excellent to poor depending on the molecular structure.

The colour fastness to washing can be improved by aftertreatment of the dyed fibre, originally by the use of after-coppering (*i.e.* treatment with copper sulphate to form a metal-complex direct dye) or by diazotisation and development. Both approaches lead to a pronounced shade change and copper in the waste water from dyeing is environmentally undesirable. Modern aftertreatments make use of cationic fixatives that may simply complex with the anionic dye, and/or form a metal complex and/or react with the cellulose fibre hydroxyl groups to form strong cova-lent bonds. Improvement in colour fastness to washing may be accompanied by some diminution in colour fastness to light.

Direct dyes may be used on all cellulosic fibres and selected dyes may also be used to dye wool, silk and nylon fibres in the manner of acid dyes. Addition of electrolyte into the dyebath increases dye aggregation in the dyebath. As the dyeing temperature is increased, the dye aggregation decreases, releasing individual direct dye anions for diffusion and adsorption inside the fibre. Direct dyes have been separated into three classes, Class A - salt-controllable, Class B - temperature-controllable, and Class C - both salt- and temperature-controllable. It is normal to select compatible dyes from within the same class, but selected dyes from Class A and B, and Class B and C may be dyed together in the same dyebath in exhaust dyeing. Direct dyes may also be applied to cellulosic fabrics by continuous pad-steam-wash off and dry methods.

The depth of colour obtained using direct dyes is deeper on viscose, lyocell and high wet modulus fibres and on mercerised cotton compared with bleached cotton, and careful dye selection is required to produce level, solid colours when dyeing blends of these fibres.

Sulphur Dyes

Sulphur dyes are chemically complex and are prepared by heating various aromatic diamines, nitrophenols, and so on with sulphur and sodium sulphide. Sulphur dyes are produced in pigment form without substantivity for cellulose. Treatment with a reducing agent (*e.g.* sodium sulphide or sodium hydrosulphide) in an alkaline dyebath converts the sulphur dye into an alkali-soluble reduced (leuco) form with substantivity for cellulosic fibres. Once absorbed within the fibre, the dye is then oxidised, usually with hydrogen peroxide, back to the insoluble pigment form. Soaping off after dyeing is important to ensure the maximum colour fastness to washing and rubbing is obtained.

Sulphur black and navy are the major dyestuffs used and the colour gamut is limited to dull colours of moderate colour fastness to washing and light. The colour fastness to chlorine and to bleach fading is poor when using multiple wash cycles with detergents containing low temperature bleach activators such as TAED (tetraacetylethylenediamine, UK) or SNOBS (sodium nonanoyloxyben-zenesulphonate, USA). Sulphur dyes cost less than many other dyes but the waste water from dyeing may require specialised treatment before release to a conventional waste water treatment plant.

Vat Dyes

Vat dyes are water-insoluble, but contain two or more keto groups (>C=O) separated by a conjugated system of double bonds that are converted into alkali-soluble enolate leuco compounds (>C—O”) by alkaline reduction, a process called vatting. The dye application method involves three stages, namely, alkaline reduction and dissolution of the vat dye (normally using sodium hydroxide and sodium dithionite (hydrosulphite)), absorption of the substantive leuco compound by the fibre, aided by electrolyte (*e.g.* sodium sulphate and wetting, dispersing and levelling agent), followed by regeneration of the vat dye inside the fibre by oxidation in air or hydrogen peroxide. The dyed fibre is then thoroughly soaped off (washed with soap or special detergent at a high temperature) to remove any surface dye and to complete any dye aggregation inside the fibre in order to obtain the final colour.

Vat dyes are based upon indanthrone, flavanthrone, pyranthrone, dibenzanthrone, acylaminoanthraquinone, carbazole, azoanthraquininone, indigoid and thioindigoid structures. Application methods include both batch exhaust dyeing and continuous pad-steam-wash off and dry methods, and apart from sodium dithionite, formaldehydesulphoxylate or hydroxyacetone may be used for the reduction stage. Vat dyes are relatively expensive but do offer outstanding colour fastness to light and washing on cellulosic fibres. In the unreduced form some vat dyes behave like disperse dyes and can therefore be applied to polyester and polyester/cellulosic blends to achieve pale-medium colour depths. For polyester/cellulosic fabrics a pad-dry-bake-chemical pad-steam-oxidise-wash off and dry production sequence is used. For exhaust dyeing, the novel technique of electrochemical reduction dyeing has been claimed to provide a redox potential of up to -960 mV suitable for the reduction of vat dyes and also sulphur dyes. Cathodic reduction of vat dyes with the addition of a mediator (a soluble reversible redox system, *e.g.* an iron [II/III]-amino complex) could potentially decrease the chemical costs for reduction and lower the chemical load in the waste water by 50-75%.

Disperse Dyes

Disperse dyes are substantially water-insoluble dyes which have substantivity for one or more hydrophobic fibres (*e.g.* secondary cellulose acetate) and are usually applied as a fine aqueous dispersion. The major chemical classes used are aminoazobenzene, anthraquinone, nitrodiphenylamine, styryl (methine), quinoph-thalone and benzodifuranone-based dyes. Disperse dyes are milled (ground) with a dispersing agent (*e.g.* polymeric forms of sodium dinaphthylmethane sulphonates) to a fine dispersion (0.5-2 |jm) and may be supplied as grains, powders or liquid dispersions.

Disperse dyes are essentially nonionic dyes that are attracted to hydrophobic fibres such as conventional polyester, cellulose triacetate, secondary cellulose acetate and nylon through non-polar forces of attraction. In the dyebath, some of the disperse dye particles dissolve to provide individual dye molecules that are small enough to diffuse into the hydrophobic fibres. The aqueous solubility of disperse dyes is low, for example 0.2-100 mgl^1 at 80 °C, but increases with increase in dyebath temperature, in turn increasing the concentration of soluble dye available for diffusion so that the dyeing rate increases.

Dispersing agents are essential because they assist in the process of decreasing the dye particle size and enable the dye to be prepared in powder and liquid forms. In addition, the dispersing agent facilitates the reverse change from powder to dispersion during dyebath preparation and maintains the dye particles in fine dispersion during dyeing. This prevents agglomeration of the dye particles in powder form and aggregation in the dyebath. The dye solubility in the dyebath can be increased by the use of levelling agents and carriers.

The general mechanism of dyeing with disperse dyes under exhaust dyeing conditions. The diffusion and adsorption of dye molecules is accompanied by desorption of some of the dye molecules from the dyed fibre back into the dyebath, to facilitate dye migration from fibre to fibre to achieve level dyeing.

The disperse dyes are considered to dye fibres via a solid solution mechanism that involves no chemical change. Each disperse dye dissolves in the fibre more or less independently of any other disperse dye present. Under constant dyebath conditions the ratio of the amount of dye on the fibre to the amount of dye in the bath is a constant at equilibrium, which varies according to the particular dye, but ultimately a separate different saturation level (or solubility limit) is achieved for each dye above which no more of that dye can be taken up by the fibre. However, most commercial dyeings are never carried out to equilibrium.

Apart from high temperature (high pressure) exhaust dyeing at 110-140 °C, disperse dyes may also be exhaust dyed using carriers (*e.g.* fibre plasticising agents that decrease the fibre T_g) at 100 °C. This method is not very popular except where blend dyeing is practised and where high temperature dyeing, if used, would degrade one of the fibres. Odourless carriers are preferred, but there are problems with carrier spotting, and carrier retention by the fibres leads to fuming of the carrier as it is volatilized off during post-heat setting treatments, while some carriers may potentally pose environmental pollution problems. Some 50% of disperse dye consumption is used for navy and black shades, and some 70% of disperse dyes are applied to polyester and polyester/ cellulose blends by exhaust dyeing.

Disperse dyes may also be applied by a pad-dry-thermofix-wash-dry process in a continuous open width fabric treatment. This utilises the sublimation properties of disperse dyes, which vaporise directly from the solid state without prior melting. Polyester can be dyed at 190-220 °C by such a process, originally called the Ther-mosol process by DuPont. Sublimation of disperse dyes is also utilised in dry-heat transfer printing, but conventional print application by printing and fixing in saturated or superheated steam is also practised, followed by washing off and drying.

To minimise potential problems linked to the presence of disperse dyes on the fibre surface after dyeing (*e.g.* poor colour fastness to washing, rubbing and tHEIR-momigration), reduction clearing after dyeing with sodium hydroxide, sodium dithionite (hydrosulphite) and a surfactant followed by hot washing is normally practised, particularly with medium-heavy depths of colour. However, some classes of disperse dyes (*e.g.* di-ester and thiophene azo dyes) are clearable by alkali alone.

Pigment Formulations

The range of chemical compounds used as pigments varies widely and includes both inorganic pigments such as titanium dioxide, and the oxides of antimony, iron and zinc, as well as carbon. Organic pigments may be based upon a very wide variety of chemical structures, for example azo, anthraquinone, dioxazine, indanthrone, isoin-dolinone, perylene, quinacridone, copper phthalocyanine, heterocyclic nickel complexes and many others. The general considerations relating to the manufacture of pigments and their application to technical textiles are separately. The finishing of pigment formulations is technically complex and strict quality control is required to ensure satisfactory application and performance of the pigment in practice.

DYES AND PIGMENTS FOR MASS COLORATION

In mass coloration, dyes or pigments are incorporated into the polymer melt (in the case of polyamide, polyester and polypropylene) or into the polymer solution (in the case of viscose, secondary cellulose acetate, cellulose triacetate and acrylic or modacrylic fibres) during fibre manufacture. The dyes that are being used pass into a dissolved phase on incorporation into the polymer melt or solution, but pigments remain as finely dispersed particles. In the main, mass pigmentation is much more widely used for colouring manufactured fibres (both synthetic and regenerated fibres) except on wet-spun acrylic fibres, where gel dyeing is used. However, mass pigmentation can be used for dry-spun acrylic polymers which are dissolved in a solvent (*e.g.* dimethyl formamide) and spun into hot air.

Technical criteria that are important for mass coloration include:

- pigment particle size and particle size distribution, both in the spinning mass and in the filament
- pigment preparations
- solubility of the colorant in the spinning mass
- colour fastness properties
- ability to remain stable under the mass processing conditions of the polymer.

Fine pigment particles (<1|jm) are required because coarse particles would interfere with the filterability of polymer solutions or melts and could impair the tensile strength of fibres, the diameter of which normally lies in the range 16-45 |jm. The pigments used must have good resistance to organic solvents and good heat stability for use in fibre manufacture.

Mass Coloration Methods

There are four main methods of incorporating colorants into manufactured fibres, which depend upon the specific fibre production process. These are:

- batch process
- injection process
- chip blending
- chip dyeing.
- Pigment preparations or solvent dyes may be used in the batch method in which the whole of the spinning mass is coloured, the colorant concentration being equal to that in the coloured filament after extrusion. In the injection process, a coloured concentrate is

continuously metered into the spinning mass which is then extruded. Chip blending is suitable only for melt coloration, for example for polyamide, polyester and polypropylene. Polymer chips are homogeneously mixed with the colour concentrate prior to extrusion. Chip dyeing is a more specialised technique used with nylon 6. The polymer chips are precoloured with polymer-soluble dyes and then melt spun.

Mass pigmentation is widely used as the major coloration route for polypropylene fibres. The flexibility of the smaller scale nature of polypropylene fibre manufacture compared with the large continuous polymerization/extrusion processes for polyamide and polyester fibres, the lower polymer melt temperature of polypropylene and the difficulties in dyeing this hydrophobic, low linear density fibre in conventional dyeing equipment, ensure that mass pigmentation is the major coloration route. Mass coloration of many other melt-spun fibres is not economic unless large fibre weights per colour may be spun for a known end-use, because of the high costs of cleaning out the production machinery to avoid subsequent colour contamination problems. However, attempts continue to find a satisfactory conventional dyeing method for modified polypropylene fibres using exhaust dyeing or continuous dyeing techniques.

Gel Dyeing of Acrylic Fibres

For wet-spun acrylic fibres, the manufacture of producer-dyed fibres involves the passage of acrylic tow in the gel state (*i.e.* never-dried state) through a bath containing basic dyes. This is utilised in the Courtelle Neochrome process (formerly Courtaulds, now Acordis) to produce dyed acrylic fibres from a continuous fibre production line typically at a speed of 50mmin^{-1} with economic batch weights per colour of 250-500 kg.

The liquid basic dyes are metered in at a rate appropriate to the acrylic tow mass and speed, the recipe being based on a computerised colour match prediction system allowing the selection of a very wide range of colours using a choice of the technically best dyes (*i.e.* easiest to apply and highest colour fastness), or lowest cost dyes, or least metameric dyes. As the freshly coagulated acrylic tow passes through the dyebath, the basic dyes diffuse inside the gel-state acrylic tow in a matter of seconds. The dyed tow is then drawn and steamed, crimped and cut to the appropriate staple length for use in technical textiles, or may be used alternatively in filament form.

Mass Coloration and Colour Fastness Properties

The colour fastness properties of manufactured fibres produced by mass dyeing, and particularly by mass pigmentation, are generally superior to those obtained on technical textiles by conventional dyeing and printing. Mass pigmentation is used where high colour fastness to light and weathering is required, for example in tenting fabrics, awnings, sun blinds, carpets, synthetic sports surfaces and so on, and where the colour can be economically produced.

The organic pigments and a few inorganic pigments used for mass pigmentation of polyester and polypropylene fibres must exhibit:

- sufficient thermal stability
- colour fastness to light, adequate for the intended end-use
- no migration (no blooming or contact bleed)
- compatibility with other additives, *e.g.* the UV stabilisers widely used in polyolefins, with no photodegradation effects on the polymer
- no adverse effect on the mechanical properties of the fibres.

High performance organic pigments are used for brilliant colours and good overall colour fastness performance on polypropylene fibres.

Inorganic pigments such as titanium dioxide (rutile), zinc oxide, and antimony oxide are used for white, and carbon (small particle size channel black) is used for black, while some iron oxide browns are used for cost reasons.

The tensile strength of mass-pigmented polypropylene fibres after 300 hours exposure in an Atlas 600 WRC accelerated light fastness testing equipment is markedly higher than the uncoloured fibre, demonstrating the protective effect of the presence of the pigments against the deleterious effects of ultraviolet radiation on the fibres.

Colour retention in awning fabrics in outdoor applications is critical, and two types of climatic conditions are most critical where pigments are incorporated within acrylic and modacrylic technical textiles:

- hot and dry (typically 38-49 °C, relative humidity <20%) with high average sunshine
- warm and humid (32-38 °C, relative humidity >70%) with average sunshine.

In hot and dry conditions, both the polymer and the pigment may degrade, whereas in warm and humid conditions pigment degradation is more likely. However, pigments that are degraded and fade under warm and humid conditions may often be quite stable under hot dry conditions.

PIGMENTS FOR TECHNICAL COATED FABRICS

Many technical textiles are coated in order to provide the high performance specification demanded by the end-use. The many considerations governing the use of coloured organic pigments in coatings, and the range of pigments available, have been reviewed by Lewis.

HIERARCHY OF TEXTILE MATERIALS

TEXTILES TECHNOLOGIES

Textiles technologies have evolved over millennia and the term 'textile' now has a very broad meaning. Originally reserved for woven fabrics, the term now applies to fibres, filaments and yarns, natural or synthetic, and most products derived from them.

This includes threads, cords, ropes and braids; woven, knitted and non-woven fabrics; hosiery, knitwear and garments; household textiles, textile furnishing and upholstery; carpets and other fibre-based floor coverings; industrial textiles, geotextiles and medical textiles.

This definition introduces three important notions. First, it states that textiles are fibrous materials. A fibre is defined as textile raw material, generally characterised by flexibility, fineness and high ratio of length to thickness; this is usually greater than 100. The diameter of fibres used in textile reinforcements for composites (glass, carbon, aramid, polypropylene, flax, etc.) varies from 5 mm to 50 mm. Continuous fibres are called *filaments*. Fibres of finite length are called *short, discontinuous*, *staple* or *chopped* with lengths from a few millimetres to a few centimetres.

Fibres are assembled into *yarns* and fibrous plies, and then into textiles. The second important feature of textiles is their hierarchical nature. One can distinguish three hierarchical levels and associated scales: (1) fibres at the *microscopic* scale; (2) yarns, repeating unit cells and plies at the *mesoscopic* scale; and (3) fabrics at the *macroscopic* scale.

Each scale is characterised by a characteristic length, say 0.01 mm for fibre diameters, 0.5–10 mm for yarn diameters and repeating unit cells, and 1–10 m and above for textiles and textile structures. Each level is also characterised by dimensionality where fibres and yarns are mostly one-dimensional while fabrics are two- or three-dimensional, and by structural organisation where fibres are twisted into yarns, yarns are woven into textiles, etc.

Textiles are structured materials. On a given hierarchical level one can think of a textile object as an entity and make abstraction of its internal structure: a yarn may be represented as a flexible rod, or a woven fabric as a membrane.

This approach is useful but the internal structure must be considered if one wishes to assess basic features and behaviour of textile objects such as the transverse compression of yarns or shear behaviour of fabrics. The diversity of textile technologies results in a large variety of available textile structures. The textile structures that are widely used as reinforcements for composites; these are discussed in this chapter.

The properties of a fabric are the properties of fibres transformed by the textile structure.

The latter is introduced deliberately during manufacturing. Modern fibres turn millennium-old textile technologies into powerful tools for creating materials designed for specific purposes, where fibre positions are optimised for each application. Textile manufacturing methods and internal structure are two important topics that are addressed in this chapter.

TEXTILE YARNS AND CLASSIFICATION

The term yarn embraces a wide range of 1D fibrous objects. A yarn has substantial length and relatively small cross-section and is made of fibres and/or filaments, with or without twist. Yarns containing only one fibre are *monofilaments*. Untwisted, thick yarns are termed *tows*. Flat tows are called *rovings* in composite parlance; in textile technology this word designates an intermediate product in spinning. *Sizing* holds the fibres together and facilitates the processing of tows; it also promotes adhesion of fibres to resin in composites. In *twisted yarns*, fibres are consolidated by the friction resulting from twist. A twist is introduced to a *continuous filament* yarn by *twisting*. For a *twisted yarn made of staple fibres* the process is called *spinning* and involves a long chain of preparatory operations.

There are different *yarn spinning* processes (*ring spinning, open-end spinning, friction spinning*) leading to yarns with different internal distributions of fibres. Note that these are distinct from fibre spinning processes such as wet spinning, melt spinning or gel spinning, which are used to make individual fibres, most often from various polymers.

Fibres of different types are easily mixed when yarn spun, producing a *blend*; thermoplastic matrix fibres can be introduced among load-carrying fibres in this way. Finally, several strand yarns can be twisted together, forming a *ply* yarn.

Linear Density, Twist, Dimensions and Fibrous Structure of Yarns

The *linear density* is the mass of a yarn per unit length; the inverse quantity is called *yarn count* or *number*. Common units for linear density. Linear density, the most important parameter of a yarn, is normalised by specifications and controlled in manufacturing. The *unevenness*, or coefficient of variation of linear density, is normally below 5% for continuous filament yarns while for spun yarns special measures in manufacturing keep it within a specified range. For yarns made of short and stiff fibres such as flax the unevenness can be as high as 15–20%.

The *twist* K of a yarn is a number of turns per unit length. Values of twist for yarns used in composite reinforcements are normally below 100 m^{-1}. The *twist angle* is the inclination of fibres on the surface of the yarn due to twist.

3D Braided Fabrics

3D braiding presents some similarities with 2D braiding. In 2D maypole braiding only two sets of carriers rotate around the braiding axis in opposite directions, creating a single textile layer.

In contrast, 3D braids can be regarded as multiple layers of interlaced yarns which are connected more or less extensively by individual yarns extending through the thickness.

It is possible to braid preforms where the level of interlacing between different layers is such that it becomes impossible to discern the distinct layers in the final preform. Carrier paths may be defined over concentric circles or Cartesian arrays, which are square or rectangular.

Two examples of 3D braiding processes are described by Byun and Chou, known as the four-step and two-step processes. The four-step process uses a framework of yarn carriers in a rectangular or circular array. As the name suggests, the process consists of four steps, each involving alternate movements of the rows and columns of yarn carriers. Between cycles the yarns are 'beaten up' into the structure and the braid is hauled off by one pitch length. The two-step process involves a large number of axial yarns arranged in the required preform geometry, with a smaller number of yarn carriers arranged around them. The carriers are moved through the array of axial yarns in two alternate directions.

Such processes can be used to produce preforms featuring yarns that extend along many directions.

Yarn directions are not limited to a plane, and this constitutes the main advantage of the process. However, any increase in out-of-plane properties is achieved at the expense of in-plane properties. 3D braids generally feature axial yarns.

Preforms of various shapes can be produced, for composite parts of which the geometry is very different to the shells associated to parts based on 2D textiles such as weaves or non-crimp fabrics. Such applications are still relatively rare; rocket engines constitute a classic example while biomedical engineering is one area of emergence.

MULTIAXIAL MULTIPLY NON-CRIMP FABRICS

The above textile reinforcements offer economical advantages while being easier to process than prepregs. However, compromises are made on the performance of the composite because of crimp that causes fibre misalignments. Non-crimp fabrics combine unidirectional crimp-free fibre layers by assembling them together by stitching – sewing or knitting – and/or bonding by chemical agents.

Terminology and Classification

A textile structure constructed out of one or more laid parallel non-crimped not-woven thread plies with the possibility of different orientations, different thread densities of single thread plies and possible integration of fibre fleeces, films, foams or other materials, fixed by loop systems or chemical binding systems.

Threads can be oriented parallel or alternating crosswise. These products can be made on machines with insertion devices (parallel-weft or cross-weft) and warp knitting machines or chemical binding systems. The definition stresses that fibrous plies are laid up with threads. Stitching the plies together by warp-knitting can be achieved in such a way that the stitches pierce the plies between the laid yarns, resulting in an open preform architecture. On the other hand wide threads (flat tows) laid close together form continuous fibrous plies bound by warp-knitting with piercing sites positioned according to the needle spacing, without any connection to tow positioning. Needles pierce the fibrous plies and distort them locally. Such a preform construction is close to ideal unidirectional uniform plies. The latter case is the most common.

Manufacturing

Producing a multiaxial multiply fabric involves laying plies or *weft insertion*, and stitching or *knitting* them together. Hence the machine has two main parts, a weft insertion device and a knitting unit.

Weft Insertion

Weft insertion carriages travelling across the machine lay several tows simultaneously at a desired angle across the fabric. The tows are fixed to the needles of the moving chains. The length of the weft insertion part of a typical machine is about 10 m and allows up to three layers with angles from 20" to 90" or –20" to –90". The speeds of weft insertion carriages dictate the fibre angles. Measurements show that the orientation of oblique fibres can deviate from the prescribed value by several degrees.

Weft inserted tows together with optional 0" plies are stitched by a warp-knitting device. A powder binder (*e.g.* epoxy) can be added beforehand to facilitate composite processing. A foil or non-woven ply may be added below the plies. A warp-knitting device as described below stitches the plies by an action of needles, which are collected in a needle bed and move upwards, piercing the plies.

Walking Needle

Stitching of ply construction by warp-knitting is optimised to decrease misalignment and disturbance of the weft yarns by the knitting needles. In older knitting units needles move in one direction only, standing at a certain angle to the plies.

The needles move linearly upwards to catch the warp knitting yarn and then down to pull the yarn through the ply, building a stitching mesh. Plies move constantly forward. The relative movement of plies and needles in the horizontal direction causes misalignment during penetration, the extent of which depends on the stitching length and other parameters. This limitation was reduced drastically by the development of the *walking needle* device, introducing a new needle movement where the needle is placed at 90" to the textile and moves in the vertical direction as well as in the direction of ply movement during penetration.

The relative horizontal motion between the textile and needle is minimal. The needle head is designed to penetrate the ply construction by pushing away the filaments of reinforcing yarns, minimising the damage.

Warp Knitting

The principle of warp knitting. An actual warp knitted fabric or a knitted web, whereby warp knitting is used to stitch unidirectional fibrous plies, is an assembly of loops formed by the interaction of needles and guides. Yarns are fed through all guides and all needles are fixed on a needle bar, moving simultaneously and forming a loop around every needle in every cycle. Consider a warp yarn about to be looped, going thorough the guide at the beginning of the knitting cycle.

Knitting needles are in their upper position. Loops forming the edge of the fabric are on the stems of the needles. The guide carries the new yarn around a needle, forming a new loop. The needle moves down, connecting the new loop with the old one sitting on its stem; in the case of ply stitching the loop is pulled through the plies. The textile is pulled and a new cycle starts. The knitting action creates loop connections in the machine (*course*) direction. Movements of the guides across the needle bed create connections in the cross (*wale*) direction, forming a warp-knit pattern. The positions of the gaps between the needles, where guides pass in subsequent knitting cycles, can be used to code the pattern by the so-called Leicester notation. Consider a diagram of the guides' movement, called a *lapping diagram*. Rows of dots represent needle positions in plane view. The numbering of needles assumes that the pattern mechanism is on the right side.

Internal Geometry

Positions of the Stitching Sites

A stitching yarn pierces the fibrous plies at positions defined by the needle spacing in the needle bed (spacing in cross direction A) and by the speed of the material feeding in the knitting device (spacing in machine direction B). The value of A is also expressed by the *machine gauge*, a number of needles per inch.

In a warp-knitted fabric devoid of fibrous plies, yarn tension leads to significant deviations of the actual loop spacing in the relaxed fabric. In the case of multiaxial multiply fabrics the stitching yarn is fixed by the fibrous plies, resulting in fairly regular spacing. An example of spacing variability for such a fabric. Deviations in the positions of stitching lead to deviations of the loops from the ideal machine direction; the deviation can be as high as 15".

Geometry of the Stitching Loop

The loop geometry can be assumed using a geometric approach or described using a mechanical approach where the shape is calculated from equilibrium equations applied to the looped yarns. The latter approach is unsuitable here as loop positions are set by the plies, and large fluctuations of loop shapes are observed in practice. Hence the geometric approach is considered more appropriate.

Stitching yarns normally have a low linear density of around 10 tex with tyically 15 filaments, and low twist ($<100\ m^{-1}$). They are easily compressible, which explains the large observed variations of their dimensions. Stitch yarn shapes are observed to vary within the repeating pattern, with cross-sections ranging from circular to flattened (elliptical) and with varying dimensions. The geometry of the yarn centreline is defined by identifying a set of anchor points A–M along the loop and approximating the centreline between the anchor points by straight lines or circular arcs.

Interactions of the Stitching with Fibrous Plies

Stitching causes deviations to fibre orientations. These deviations can result in linear channels or be localised near stitching sites as *cracks*. Localised cracks have a rhomboidal shape of width b and length l. A channel is formed when localised cracks touch or overlap. The direction of cracks and channels corresponds to that of fibres in the ply. The cracks/channels are also evident inside the fabric, where they provide routes for resin flow. In a composite they create resin rich zones which can play an important role in the initiation of damage.

Crack and channel dimensions feature much scatter. The measured crack and channel dimensions for various fabrics. Crack widths are approximately proportional to the thickness of the stitch yarn, so that an empirical coefficient k can be used to calculate $b = kd_0$, where d_0 is the compacted stitch diameter. Fibres in a powdered fabric are more difficult to displace, explaining the low values of k and l/b for the latter fabric. In non-powdered fabrics channels are wider than cracks, with rough k averages of 4 (cracks) and 7 (channels) and l/w values for cracks of about 20. Dimensions are also influenced by knitting tension.

When large channels extend along the machine direction, stitch yarn sections on the fabric face or loops on the back can sink in the channels. This phenomenon plays a major role in nesting of non-crimp fabric layers in laminates. When this sinking occurs the layers can come closer, closing any gap between them that might be introduced by stitched yarns laying on the surface of the fabric. In the former, stitch yarns extend across the fibre direction and cannot sink in cracks. The face of the latter textile has wide channels into which stitches sink deeply, allowing nesting. Hence the thickness per layer of the compressed fabric 'Q' decreases with an increase in number of layers while it stays constant for fabric 'B'.

MODELLING OF INTERNAL GEOMETRY OF TEXTILE PREFORMS

Textiles are hierarchically structured fibrous materials. As discussed in the classic paper by Hearle *et al*. this description of the nature of textiles allows efficient construction of mathematical models for the geometry and the mechanical behaviour of textile structures. In spite of the generally recognised usefulness of this approach, it has not been used to its full strength for the creation of textile structure models. During the 1930s the first serious mechanical treatments of the structure of textile materials were published by Peirce, Pozdnyakov and Novikov. Since then, numerous publications have been proposed on the mechanical behaviour of textiles, culminating in a comprehensive treatment on *Mechanics of Flexible Fibre Assemblies* in 1980. In the following years the ideas and approaches outlined in this book were pursued further. Textile mechanics at the outset of the 21st century includes models of the internal geometry of textile structures such as continuous-filament and staple yarns, random fibre mats, woven and knitted fabrics. The hierarchical description of textile structures is implemented using the minimum energy principle as introduced by Hearle and Shanahan and de Jong and Postle. This allows the decomposition of a structure into a set of structural elements, leading to models that are physically sound and computationally feasible. It should be noted that the principle of minimum energy is heuristic when applied to non-conservative mechanical systems such as textiles.

Following the hierarchical approach one can consider a description of the internal geometry of a fabric on two different levels: the *yarn paths mode* describing the spatial configuration of yarns and the *fibre distribution mode* describing spatial distribution properties of the fibrous assembly. In the former case the fabric is regarded as an agglomeration of yarns – slender curved bodies. Their spatial positions are defined by description of yarn midlines and cross-section at each point of the midlines. The internal fibrous structure of the yarns is not considered. In the latter case the fibrous structure of yarns should be defined. The definition of the spatial positions of yarns states whether any point in the unit cell lies inside a yarn or not; if so, parameters of the fibrous assembly at this point are determined. The yarn path mode is sufficient for the calculation of simpler data related to the internal geometry, such as surface density, fabric thickness and local inter-yarn porosity. The fibre distribution mode is needed when detailed information on the fibrous structure is required.

3

Tools Used in Garment Construction

Besides a sewing machine in good condition, well selected sewing equipment and pressing equipment are essential for making garments of good quality and appearance. Sewing box/ kit: First and foremost, one should have a box or basket with compartments for keeping the things needed for sewing.

HISTORY OF CLOTHING AND TEXTILES

The wearing of clothing is exclusively a human characteristic and is a feature of most human societies. It is not known when humans began wearing clothes. Anthropologists believe that animal skins and vegetation were adapted into coverings as protection from cold, heat and rain, especially as humans migrated to new climates; alternatively, covering may have been invented first for other purposes, such as magic, decoration, cult, or prestige, and later found to be practical as well.

Clothing and textiles have been important in human history and reflects the materials available to a civilization as well as the technologies that it has mastered. The social significance of the finished product reflects their culture.

Textiles, defined as felt or spun fibers made into yarn and subsequently netted, looped, knit or woven to make fabrics, appeared in the Middle East during the late stone age. From ancient times to the present day, methods of textile production have continually evolved, and the choices of textiles available have influenced how people carried their possessions, clothed themselves, and decorated their surroundings.

Sources available for the study of the history of clothing and textiles include material remains discovered via archaeology; representation of textiles and their manufacture in art; and documents concerning the manufacture, acquisition, use, and trade of fabrics, tools, and finished garments. Scholarship of textile history, especially its earlier stages, is part of material culture studies.

Prehistoric Development

Modern humans are the only survivors of several species of naked apes who may have worn clothes, according to DNA studies of clothing lice. This study suggests that clothing may possibly have been used 650 thousand years ago – much longer than previously thought. First fabric uses, likely to be felt, are thought to have been used about 100,000 years ago.

Early Adoption of Fibrous Apparel

Another genetic analysis suggests that the human body louse, which lives in clothing, may only have diverged from the head louse some 107 thousand years ago, which supports evidence that humans began wearing clothing at around this time. These estimates pre-date the first known human exodus from Africa, although species of Homo (other than Homo Sapiens) who may have worn clothes - and shared these louse infestations - appear to have migrated earlier.

Initial Manufacture of Clothes

The development of textile and clothing manufacture in prehistory has been the subject of a number of scholarly studies since the late 20th century. These sources have helped to provide a coherent history of these prehistoric developments. Evidence suggests that human beings may have begun wearing clothing as far back as 100,000 to 500,000 years ago. Possible sewing needles have been dated to around 40,000 years ago. The earliest definite examples of needles originate from the Solutrean culture, which existed in France from 19,000 BC to 15,000 BC. The earliest dyed flax fibers have been found in a prehistoric cave in the Republic of Georgia and date back to 36,000 BP.

The earliest evidence of weaving comes from impressions of textiles and basketry and nets on little pieces of hard clay, dating from 27,000 years ago and found in Dolni Vestonice in the Czech Republic. At a slightly later date (25,000 years) the Venus figurines were depicted with clothing. Those from western Europe were adorned with basket hats or caps, belts worn at the waist, and a strap of cloth that wrapped around the body right above the breast. Eastern European figurines wore belts, hung low on the hips and sometimes string skirts. Archaeologists have discovered artifacts from the same period that appear to have been used in the textile arts: (5000 BC) net gauges, spindle needles and weaving sticks.

Ancient Textiles and Clothing

The first actual textile, as opposed to skins sewn together, was probably felt. Surviving examples of Nålebinding, another early textile method, date from 6500 BC. Our knowledge of ancient textiles and clothing has expanded in the recent past thanks to modern technological developments. Our knowledge of cultures varies greatly with the climatic conditions to which archeological deposits are exposed; the Middle East and the arid fringes of China have provided many very early samples in good condition, but the early development of textiles in the Indian subcontinent, sub-Saharan Africa and other moist parts of the world remains unclear. In northern Eurasia peat bogs can also preserve textiles very well. Early woven clothing was often made of full loom widths draped, tied, or pinned in place.

Ancient Near East

The earliest known woven textiles of the Near East may be fabrics used to wrap the dead, excavated at a Neolithic site at Çatalhöyük in Anatolia, carbonized in a fire and radiocarbon dated to c. 6000 BC. Evidence exists of flax cultivation from c. 8000 BC in the Near East, but the breeding of sheep with a wooly fleece rather than hair occurs much later, c. 3000 BC.

Ancient India

The inhabitants of the Indus Valley Civilization used cotton for clothing as early as the 5th millennium BC – 4th millennium BC. According to The Columbia Encyclopedia, Sixth Edition: "Cotton has been spun, woven, and dyed since prehistoric times. It clothed the people of ancient India, Egypt, and China. Hundreds of years before the Christian era cotton textiles were woven in India with matchless skill, and their use spread to the Mediterranean countries. In the 1st cent. Arab traders brought fine Muslin and Calico to Italy and Spain. TheMoors introduced the cultivation of cotton into Spain in the 9th cent. Fustians and dimities were woven there and in the 14th cent. in Venice and Milan, at first with a linen warp. Little cotton cloth was imported to England before the 15th cent., although small amounts were obtained chiefly for candlewicks. By the 17th cent. the East India Company was bringing rare fabrics from India. Native Americans skillfully spun and wove cotton into fine garments and dyed tapestries. Cotton fabrics found in Peruvian tombs are said to belong to a pre-Inca culture. In colour and texture the ancient Peruvian and Mexican textiles resemble those found in Egyptian tombs."

Ancient Egypt

Evidence exists for production of linen cloth in Ancient Egypt in the Neolithic period, c. 5500 BC. Cultivation of domesticated wild flax, probably an import from the Levant, is documented as early as c. 6000 BC Other bast fibers including rush, reed, palm, and papyrus were used alone or with linen to make rope and other textiles. Evidence for wool production in Egypt is scanty at this period. Spinning techniques included the drop spindle, hand-to-hand spinning, and rolling on the thigh; yarn was also spliced. A horizontal ground loom was used prior to the New Kingdom, when a vertical two-beam loom was introduced, probably from Asia. Linen bandages were used in the burial custom of mummification, and art depicts Egyptian men wearing linen kilts and women in narrow dresses with various forms of shirts and jackets, often of sheer pleated fabric.

Ancient China

The earliest evidence of silk production in China was found at the sites of Yangshao culture in Xia, Shanxi, where a cocoon of bombyx mori, the domesticated silkworm, cut in half by a sharp knife is dated to between 5000 and 3000 BC. Fragments of primitive looms are also seen from the sites ofHemudu culture in Yuyao, Zhejiang, dated to about 4000 BC. Scraps of silk were found in a Liangzhu culture site at Qianshanyang in Huzhou, Zhejiang, dating back to 2700 BC. Other fragments have been recovered from royal tombs in the Shang Dynasty (c. 1600 – c. 1046 BC). Under the Shang Dynasty, Han Chinese clothing or Hanfu consisted of a *yi*, a narrow-cuffed, knee-length tunic tied with a sash, and a narrow, ankle-length skirt, called *shang*, worn with a *bixi*, a length of fabric that reached the knees. Clothing of the elite was made of silk in vivid primary colours.

Ancient Japan

The earliest evidence of weaving in Japan is associated with the Jômon period. This culture is defined by pottery decorated with cord patterns. In a shell mound in the Miyagi Prefecture, dating back about 5,500, some cloth fragments were discovered made from bark fibers. Hemp fibers were also discovered in the Torihama shell midden, Fukui Prefecture, dating back to the Jômon period, suggesting that these plants could also have been used for clothing. Some pottery pattern imprints depict also fine mat designs, proving their waving techniques. Since bone needles were also found, it is assumed that they wore dresses that were sewn together.

THE TEXTILE TRADE IN THE ANCIENT WORLD

The exchange of luxury textiles was predominant on the Silk Road, a series of ancient trade and cultural transmission routes that were central to cultural interaction through regions of the Asian continent connecting East and West by linking traders, merchants, pilgrims, monks, soldiers, nomads and urban dwellers from China to the Mediterranean Sea during various periods of time. The trade route was initiated around 114 BC by the Han Dynasty, although earlier trade across the continents had already existed. Geographically, the Silk Road or Silk Route is an interconnected series of ancient trade routes between Chang'an in China, with Asia Minor and the Mediterranean extending over 8,000 km (5,000 mi) on land and sea. Trade on the Silk Road was a significant factor in the development of the great civilizations of China, Egypt, Mesopotamia, Persia, the Indian subcontinent, and Rome, and helped to lay the foundations for the modern world.

Classical Antiquity

Dress in classical antiquity favored wide, unsewn lengths of fabric, pinned and draped to the body in various ways. Ancient Greek clothing consisted of lengths of wool or linen, generally rectangular and secured at the shoulders with ornamented pins called fibulae and belted with a sash. Typical garments were the peplos, a loose robe worn by women; the chlamys, a cloak worn by men; and the chiton, a tunic worn by both men and women. Men's chitons hung to the knees, whereas women's chitons fell to their ankles. A long cloak called a himation was worn over the peplos or chlamys.

The toga of ancient Rome was also an unsewn length of wool cloth, worn by male citizens draped around the body in various fashions, over a simple tunic. Early tunics were two simple rectangles joined at the shoulders and sides; later tunics had sewn sleeves. Women wore the draped stola or an ankle-length tunic, with a shawl-like palla as an outer garment. Wool was the preferred fabic, although linen, hemp, and small amounts of expensive imported silk and cotton were also worn.

Iron Age Europe

The Iron Age is broadly identified as stretching from the end of the Bronze Age around 1200 BC to 500 AD and the beginning of the Medieval period. Bodies and clothing have been found from this period, preserved by the anaerobic and acidic conditions of peat bogs in northwestern Europe. A Danish recreation of clothing found with such bodies indicates woven wool dresses, tunics and skirts.

These were largely unshaped and held in place with leather belts and metal brooches or pins. Garments were not always plain, but incorporated decoration with contrasting colours, particularly at the ends and edges of the garment.

Men wore breeches, possibly with lower legs wrapped for protection, although Boucher states that long trousers have also been found.Warmth came from woollen shawls and capes of animal skin, probably worn with the fur facing inwards for added comfort. Caps were worn, also made from skins, and there was an emphasis on hair arrangements, from braids to elaborate Suebian knots. Soft laced shoes made from leather protected the foot.

Medieval Clothing and Textiles

The history of Medieval European clothing and textiles has inspired a good deal of scholarly interest in the 21st century. Elisabeth Crowfoot, Frances Pritchard, and Kay Staniland authored*Textiles and Clothing: Medieval Finds from Excavations in London, c.1150-c.1450* (Boydell Press, 2001). The topic is also the subject of an annual series *Medieval Clothing and Textiles* (Boydell Press) edited by Robin Netherton and Professor Gale R. Owen-Crocker of Anglo-Saxon Culture at the University of Manchester.

Byzantium

The Byzantines made and exported very richly patterned cloth, woven and embroidered for the upper classes, and resist-dyed and printed for the lower. By Justinian's time the Roman toga had been replaced by the tunica, or long *chiton*, for both sexes, over which the upper classes wore various other garments, like a *dalmatica* (dalmatic), a heavier and shorter type of tunica; short and long cloaks were fastened on the right shoulder.

Leggings and hose were often worn, but are not prominent in depictions of the wealthy; they were associated with barbarians, whether European or Persian.

Early Medieval Europe

European dress changed gradually in the years 400 to 1100. People in many countries dressed differently depending on whether they identified with the old Romanised population, or the new invading populations such as Franks, Anglo-Saxons, and Visigoths. Men of the invading peoples generally wore short tunics, with belts, and visible trousers, hose or leggings. The Romanised populations, and the Church, remained faithful to the longer tunics of Roman formal costume.

The elite imported silk cloth from the Byzantine, and later Muslim worlds, and also probably cotton. They also could afford bleached linen and dyed and simply patterned wool woven in Europe itself. But embroidered decoration was probably very widespread, though not usually detectable in art. Lower classes wore local or homespun wool, often undyed, trimmed with bands of decoration, variously embroidery, tablet-woven bands, or colorful borders woven into the fabric in the loom.

High Middle Ages and the Rise of Fashion

Clothing in 12th and 13th century Europe remained very simple for both men and women, and quite uniform across the subcontinent. The traditional combination of short tunic with hose for working-class men and long tunic with overgown for women and upper class men remained the norm. Most clothing, especially outside the wealthier classes, remained little changed from three or four centuries earlier.

The 13th century saw great progress in the dyeing and working of wool, which was by far the most important material for outer wear. Linen was increasingly used for clothing that was directly in contact with the skin. Unlike wool, linen could be laundered and bleached in the sun. Cotton, imported raw from Egypt and elsewhere, was used for padding and quilting, and cloths such as buckram and fustian.

Crusaders returning from the Levant brought knowledge of its fine textiles, including light silks, to Western Europe. In Northern Europe, silk was an imported and very expensive luxury. The well-off could afford woven brocades from Italy or even further afield. Fashionable Italian silks of this period featured repeating patterns of roundels and animals, deriving from Ottoman silk-weaving centres in Bursa, and ultimately from Yuan Dynasty China via the Silk Road.

Cultural and costume historians agree that the mid-14th century marks the emergence of recognizable "fashion" in Europe. From this century onwards Western fashion changes at a pace quite unknown to other civilizations, whether ancient or contemporary. In most other cultures only major political changes, such as the Muslim conquest of India, produced radical changes in clothing, and in China, Japan, and the Ottoman Empire fashion changed only slightly over periods of several centuries.

In this period the draped garments and straight seams of previous centuries were replaced by curved seams and the beginnings of tailoring, which allowed clothing to more closely fit the human form, as did the use of lacing and buttons. A fashion for *mi-parti* or *parti-coloured* garments made of two contrasting fabrics, one on each side, arose for men in mid-century, and was especially popular at the English court. Sometimes just the hose would be different colours on each leg.

RENAISSANCE AND EARLY MODERN PERIOD

Renaissance Europe

Wool remained the most popular fabric for all classes, followed by linen and hemp. Wool fabrics were available in a wide range of qualities, from rough undyed cloth to fine, dense broadcloth with a velvety nap; high-value broadcloth was a backbone of the English economy and was exported throughout Europe. Wool fabrics were dyed in rich colours, notably reds, greens, golds, and blues.

Silk-weaving was well established around the Mediterranean by the beginning of the 15th century, and figured silks, often silk velvets with silver-gilt wefts, are increasingly seen in Italian dress and in the dress of the wealthy throughout Europe. Stately floral designs featuring a pomegranate or artichoke motif had reached Europe from China in the previous century and became a dominant design in the Ottoman silk-producing cities of Istanbul and Bursa, and spread to silk weavers inFlorence, Genoa, Venice, Valencia and Seville in this period.

As prosperity grew in the 15th century, the urban middle classes, including skilled workers, began to wear more complex clothes that followed, at a distance, the fashions set by the elites. National variations in clothing increased over the century.

Early Modern Europe

By the first half of the 16th century, the clothing of the Low Countries, German states, and Scandinavia had developed in a different direction than that of England, France, and Italy, although all absorbed the sobering and formal influence of Spanish dress after the mid-1520s.

Elaborate slashing was popular, especially in Germany. Black was increasingly worn for the most formal occasions. Bobbin lacearose from passementerie in the mid-16th century, probably in Flanders. This century also saw the rise of the ruff, which grew from a mere ruffle at the neckline of the shirt or chemise to immense cartwheel shapes. At their most extravagant, ruffs required wire supports and were made of fine Italian reticella, a cutwork linen lace. By the turn of the 17th century, a sharp distinction could be seen between the sober fashions favored by Protestants in England and the Netherlands, which still showed heavy Spanish influence, and the light, revealing fashions of the French and Italian courts.

The great flowering of needlelace occurred in this period. Geometric reticella deriving from cutwork was elaborated into true needlelace or *punto in aria* (called in England "point lace"), which reflected the scrolling floral designs popular for embroidery. Lacemaking centres were established in France to reduce the outflow of cash to Italy. According to Dr. Wolf D. Fuhrig, "By the second half of the 17th century, Silesia had become an important economic pillar of the Habsburg monarchy, largely on the strength of its textile industry."

Enlightenment and the Colonial Period

During the eighteenth century, distinction was made between *full dress* worn at Court and for formal occasions, and *undress* or everyday, daytime clothes. As the decades progressed, fewer and fewer occasions called for full dress which had all but disappeared by the end of the century. Full dress followed the styles of the French court, where rich silks and elaborate embroidery reigned. Men continued to wear the coat, waistcoat and breeches for both full dress and undress; these were now sometimes made of the same fabric and trim, signalling the birth of the three-piece suit.

Women's silhouettes featured small, domed hoops in the 1730s and early 1740s, which were displaced for formal court wear by side hoops or panniers which later widened to as much as three feet to either side at the court of Marie Antoinette. Fashion reached heights of fantasy and abundant ornamentation, before new enthusiasms for outdoor sports and country pursuits and a long-simmering movement toward simplicity and democratization of dress under the influence of Jean-Jacques Rousseau and the American Revolution led to an entirely new mode and the triumph of British woollen tailoring following the French Revolution.

For women's dresses, Indian cottons, especially printed chintzes, were imported to Europe in large numbers, and towards the end of the period simple white muslin gowns were in fashion.

Industrial Revolution

During the industrial revolution, fabric production was mechanised with machines powered by waterwheels and steam-engines. Production shifted from small cottage based production to mass production based on assembly line organisation. Clothing production, on the other hand, continued to be made by hand. Sewing machines emerged in the 19th century streamlining clothing production. In the early 20th century workers in the clothing and textile industries became unionised. Later in the 20th century, the industry had expanded to such a degree that such educational institutions as UC Davis established a Division of Textiles and Clothing, The University of Nebraska-Lincoln also created a Department of Textiles, Clothing and Design that offers a Masters of Arts in Textile History, and Iowa State University established a Department of Textiles and Clothing that features a History of costume collection, 1865–1948. Even high school libraries have collections on the history of clothing and textiles.

Alongside these developments were changes in the types and style of clothing produced. During the 1960s, had a major influence on subsequent developments in the industry. Textiles were not only made in factories. Before this that they were made in local and national markets. Dramatic change in transportation throughout the nation is one source that encouraged the use of factories. New advances such as steamboats, canals, and railroads lowered shipping costs which caused people to buy cheap goods that were produced in other places instead of more expensive goods that were produced locally. Between 1810 and 1840 the development of a national market prompted manufacturing which tripled the output's worth. This increase in production created a change in industrial methods, such as the use of factories instead of hand made woven materials that families usually made.

The vast majority of the people who worked in the factories were women. Women went to work in textile factories for a number of reasons. Some women left home to live on their own because of crowding at home; or to save for future marriage portions.

The work enabled them to see more of the world, to earn something in anticipation of marriage, and to ease the crowding within the home. They also did it to make money for family back home. The money they sent home was to help out with the trouble some of the farmers were having. They also worked in the millhouses because they could gain a sense of independence and growth as a personal goal.

SEWING AND EMBROIDERY TOOLS

Hand Sewing Needles

They are found in sizes from the very fine 9 to the heavy 18. The best quality needles are made of hand ground steel. For hand sewing medium length needles with a short oval eye is selected. Crewel needles designed for embroidery work have a long oval eye.

Sewing Machine Needles

They are found in sizes from the fine 9 to the heavy 18. The needles are made to fit the specific make and model of each sewing machine. The needle size should conform to the weight, thickness, and kind of fabric.

Sewing Thread

With the wide and ever increasing range of fabrics available in the market, it is important to know the right sewing thread for the various types of fabrics. The right kind of thread is important in sewing as both the thread and the garment should share the same characteristics, as they have to be laundered and ironed together, they should shrink and stretch together.

Pins

These come in different sizes for use in different fabrics. The right choice of pins is most essential for good workmanship, speed, and convenience in sewing. Use silk or stainless steel pins. The ball point pins are useful for fine knits. The other types of pins are dressmaker pin (a pin of medium diameter but quite suitable for most sewing needs),and silk (a very slender pin with a needle point to be used on delicate fabrics).

Thimbles

These are necessary for efficient and accurate hand sewing. A metal thimble should fit snugly on the middle finger of the needle holding hand. There are two types of thimbles: an open-ended thimble, preferred by tailors, and the more common closed-ended thimble, called the dressmakers thimble.

This is used for keeping the fabric stretched while the work is being carried out.

Embroidery Threads

These are available in a variety of colours in six stranded skeins. One can use two or three strands at a time.

Stiletto

This is a sharp pointed instrument for punching holes in material. It is used for forming holes in material. It is used for forming eyelets in belts and for embroidery work.

Bodkin

This is a flat needle with a blunt end and a large eye for threading elastic and tape. The following types of shears and scissors are made for both right-handed and left- handed cutting. All cutting tools must be kept sharp, clean, and grease-free for accurate cuts.

CUTTING TOOLS

They are 8 to 10 inches long. They are used for cutting all types of fabrics. Shears differ from scissors in that they have one small ring handle for the thumb and a large ring handle for the second, third and fourth fingers.

Scissors

They are 5 to 6 inches long. They are used for light cutting, trimming, clipping corners, and cutting curves. These have round handles for both the blades. They are designed for snipping threads and trimming seams. They should be held so that the wider blade is above the narrower blade.

Pinking Shears

They are 9 to 10 inches long. They produce a notched cutting line (zig zag) which gives a neat appearance to the inside of garments.

TOOLS USED IN GARMENT CONSTRUCTION

Button Hole Scissors

These can be adjusted so as to cut button holes in any size required. They are useful if one needs to make many button holes.

Electric Scissors

These are used in most sample rooms. They are ideal for cutting silk, nylon, and soft, hard-to-cut fabric.

MEASURING TOOLS

Measuring Tape

It has a smooth surface that is clearly marked with increments of inches and centimeters on both sides. It is usually about ½ inch to ¼ inch wide and 60 inches long, and has 1/8 divisions. At one end of the tape is attached a brass strip about 3 inches long and at the other end, a small brass covering.

Rulers

They are used in sample room which are either clear plastic or metal. It is useful to have 2 rulers: one is 1 inch wide and 6 inches long, and the second is 2 inches wide and 18 inches long.

Yardstick or Meter Scale

It is available in 36 inches or 45 inches in length in wood or metal. They are useful for checking grain lines when fixing pattern pieces on material and for drawing long seam lines on fabric or paper.

L Square

It is an L-shaped metal ruler; the long arm measures 24 inches, the short arm 14 inches. This has a perfect right angle corner and is used to draw lines at right angles at the time of drafting. It is helpful during the process of "straightening fabric" to check whether the corners of the fabric have got the right-angled structure

Skirt Marker

They mark hem length accurately. Markers are adjustable and are available for use with chalk powder or pins. A six-inch gauge can be made of cardboard or bought from a shop. It is useful as a measuring guide for marking width of hems, pleats, seam allowances etc. accurately. Notches are provided at regular intervals along the gauge.

One edge of the notch is at right angles to the straight. While measuring or marking, use the straight edge of the notch as a guide.

GARMENT CONSTRUCTION AND TOOLS

Tracing Wheel

It is used to transfer the pattern markings-including seams, darts and pocket placements to the wrong side of the fabric with the aid of tracing paper. The small serrated edge tracing wheel is appropriate for most fabrics. A smooth edge tracing wheel is used on fine or knit fabrics to avoid snagging the yarns.

Tracing Paper

It is a wax-coated paper used with the tracing wheel to transfer pattern markings to the wrong side of the fabric

Tailors Chalk

It is made of wax or stone chalk that is used to transfer markings to fabric when white carbon is not visible. Stone chalk is also available in pencil form. This is available in assorted colours and in rectangular or triangular shapes.

PRESSING TOOLS

Iron

This keep an automatic iron handy for pressing fabric before cutting, during construction and after the garment is completed.

Steam Iron

It has an adjustable temperature control, and is equipped with a thumb press for automatic steam. Distilled water is heated, and the resulting steam can be released with the thumb press while pressing.

Ironing Board

It is used for hand pressing which is padded and of convenient height. One may use an ordinary table covered with sheet and blanket for this purpose.

Sleeve Board

This is a well-padded miniature of a full-sized ironing board and is used to press sleeves and hard-to-reach small.

It has a tapered end on one side and a round end on the other side.

Press Cloth

Fabric used for press cloth should be colourfast and should be washed or boiled to remove all sizing.

MISCELLANEOUS TOOLS

Awl

It is a small, sharp-pointed tool used to punch small, round holes for marking on paper or leathers.

Seam Ripper

It is used to remove basting stitches and to rip out unnecessary machine stitches.

Loop Turner

Is a long wire with a latch hook, used for turning bias strips to make spaghetti straps and narrow belts.

Dress Form

This is a padded form of body and may be made of wood, cardboard, plaster of paris, reinforced plastic. This is a necessity in all sample rooms for designing and fittings.

Orange-stick

This is a long tool whose point can be inserted into the corners of collars, seams, etc., so as to give a neat pointed appearance. Cutting board or table:

This is a table of convenient height and size is a definite aid in cutting and constructing garments. 5' x 3' is a desirable size with 2" x 6" height.

ACTION OF THE SEWING MECHANISM

The action of the sewing mechanism. Just before the stitching action takes place, the driving belt is on the loose pulley and is imparting motion to the pulley *be* by means of the pulleys b_3, b_t, $b_{s>}$. When the ends of the cuts have been placed on the pins and are in a suitable position to be stitched, the rod *d** is moved to the left, which throws the leather band from the loose pulley *b.* to the tight pulley *b,* and imparts motion to the pulley *b,* and the band *c.* This band passes around the two binder pulleys c_u *c,* and a pulley c,, on the sewing-machine head, and by this means starts the stitching mechanism in operation.

At the same time, the sewing-machine head travels from one side of the machine to the other. The mechanism that accomplishes this part of the operation is as follows: the sewing-machine head imparts motion to the worm c_s, which drives the worm-gear c_t. The worm-gear *c* forms a friction clutch with the casting $<r_5$, which carries a key projecting into a keyway in the shaft *c.* At the upper end of this shaft is a thread on which is a hand wheel *c,,* while at its lower end is a gear *c,,* the teeth of which work in a stationary screw *c.* Before starting the stitching mechanism the hand wheel *c,* is screwed down, which forces the casting c_s into contact with the worm-gear c_4; consequently, as the worm *c,* imparts motion to the worm-gear c_t, the shaft *c,* is rotated, thus giving motion to the gear *c.* Since the screw *c,* is stationary, the gear *c,,* as it revolves, must necessarily receive a horizontal motion. It is by this means that the sewing-machine head is moved automatically from one side of the machine to the other during the stitching operation.

The plate d_t on which the sewing-machine head rests, slides on the supports d_s, d_e and carries a projection *d,* which, when the head has traveled the full width of the cloth, comes in contact with a casting setscrewed to the rod *d,;* this moves the rod *d,* to the right and shifts the band from the tight pulley *b,* to the loose pulley *b,,* thus stopping automatically the stitching mechanism when the seam has been completed. To move the stitching mechanism back to its original position the hand wheel c_6 is unscrewed, which allows the springs c_{la} to force the casting *c,* out of contact with the worm-gear *c* and breaks the connection between this gear and the shaft *c* When the parts are in this position, the plate d_t together with the different parts of its supports, may be moved by hand to any desired position.

After the two ends have been stitched, they are removed from the pins d_{lt} *d** and are allowed to fall between the rods *a,,* d_3. The rolling mechanism is then set in operation by pressing down the foot-lever *b,,* when the second cut of cloth is unrolled and rewound on the roll at the back of the machine. This operation is repeated with a sufficient number of cuts to form one large roll, which should be made large enough to facilitate the operation at the future machines, but not so large that it cannot be readily handled. Rolls are ordinarily made from 28 inches to 32 inches in diameter and contain from 800 yards to 1,200 yards of cloth.

The plate *d* to which the supports for the handle *d,* and the pins d_x are attached, is setscrewed to the supports *d,, d,,* and does not move during the stitching operation.

The plate d_e is not attached to the plate d_t, although the latter is shown close to the former, as the sewing machine is there represented in its initial, or starting, position with the presser foot ready to pass between the pins d_x as the plate d_t and sewing machine start on their movement to the right. On the other side of the machine is a corresponding plate *d* carrying a pillar rf_{10} supporting the pins d_a. The plate *d,* is capable of sliding on the supports d_s, *d,,* but does not move during the stitching operation, being secured to the supports by means of a thumbscrew d_{llt}.

The guide plates a_3, *a,,* shown attached to the rod are secured to the rod by thumbscrews in such a position that they serve to guide the cloth correctly while it is being wound on the roll at the back of the machine, and also in suitable relative position to the pins d_{lt} d_a. It is not customary to change the positions of the plate *d,,* the pins d_t, and plate *a,* except in extreme cases. When it is desired to make adjustments for different widths of cloth, the thumbscrew rfn, is loosened and the plate *d,* moved inwards or outwards to bring the pins *d,* into the desired position, after which the thumbscrew is again secured. At the same time the guide plate a_A attached to the rod *a,,* is adjusted.

The sewing-and-rolling machine that has been described occupies a floor space of about 7 feet by 4 feet. The driving pulleys are 12 inches by 2| inches, and are usually operated at a speed of 320 revolutions per minute. When operated at its full capacity, it can sew and wind in a day from 500 to 600 cuts, averaging 50 yards in length. In some cloth rooms, the sewing machine is independent of the rolling attachment. In case a stationary machine is desired, that portion of the machine shown in the upper part is mounted on high stands resting on the floor and is driven by hand or by power as desired. Another arrangement when the sewing attachment is required to be used in connection with some machine, other than a rolling machine, is to secure the upper portion of the mechanism to short stands suitable for attachment to any machine desired.

PORTABLE SEWING MACHINES

16. In many cases, it is necessary to stitch the ends of two pieces of cloth together in some part of the cloth room where it is more convenient to take the sewing machine to the cloth than to bring the cloth to the machine. This is the case when the stitches that bind the ends of two pieces together in a large roll have been broken or cut, through some cause, and a part at least of the two ends separated. It would be undesirable to run the entire roll of cloth through the sewing-and-rolling machine simply for the purpose of putting in these few stitches, and, in fact, in many cases it would not be possible, as the defective stitches are often not discovered until the cloth is being run through another machine.

Sometimes it is desirable to have one sewing machine available for use with several other machines; for example, when the cloth from a large roll has been run through a later machine and it is desired to attach the end of a new roll to the tail-end of the one that has been run through. In such cases, a portable foot-power sewing machine.

This machine rests on wheels and consequently can be moved readily from one part of the room to another, as may be desired. In operation the ends of the two pieces of cloth that are to be sewn together are attached by pins to a slowly rotating wheel. As the machine is operated by means of the treadle, this wheel feeds the cloth under the presser foot of the sewing attachment where the stitching is inserted. In other cases where the amount of available floor space is limited, the sewing machine is suspended from the ceiling of the room, and in still other cases where one sewing machine is required at various points in the room, the sewing mechanism is suspended from a track attached to the ceiling, so that it may be moved to different points in the room where it is required. In such cases, of course, the sewing is performed by hand power.

ROLLING MACHINES AND HAND SEWING

Many times, especially in the older cotton mills, machines are used for the same purpose as the sewing-and-rolling machine described, but constructed and operated without the sewing attachment. In general construction, they are similar to the machine, although for special fabrics, such as those exceptionally wide and heavy, different styles of construction are adopted. The mechanism of the machine requires no description beyond that already given.

With these machines, the ends of consecutive pieces of cloth are sewed together by hand, the operator using for this purpose a long, heavy needle with thin twine knotted at one end and longer than the width of the fabric. The tail-end of each piece is placed against the new end from another piece and the needle passed alternately through the cloth from one side to the other at intervals of about 2 inches. The knotted end of the twine remains at one edge of the cloth, while the operator ties another knot at the other edge after the thread has been passed through the cloth, so as to prevent its being drawn out in passing on to the roll or through a succeeding machine. This system is largely used in cloth rooms where consecutive pieces of cloth are sewed together temporarily, since it occupies less time and the thread can be easily drawn out later, but is not as satisfactory as machine sewing.

EQUIPMENT AND THE OPERATION OF THE CLOTH ROOM

The equipment and the operation of the cloth room cannot receive too much attention; in too many mills it is given but little consideration. In the designing or the equipment of the mill, very frequently any available space is utilized for the cloth room, while after the mill is in operation, so long as the goods are inspected and baled, and the room is operated at a minimum expense, it receives no further attention. As a matter of fact, the proper equipment and operation of the cloth room is of vital importance to the successful running of the mill. The construction and location of the cloth room should be of the best, its equipment the most suitable for the fabrics being manufactured, and the arrangement of the machinery such as will enable the cloth to be treated with the least amount of handling and the greatest economy. The machines used should be so selected, set, and operated as to effect an improvement in the cloth between its leaving the loom and being packed into bales, that will far more than compensate for the comparatively small expense per yard involved. This will increase the reputation of the mill for producing well-made, well-classified, and well-finished fabrics, and will result ultimately in increased prices or preference when orders are being placed. Careful and constant attention to this department cannot fail to result in creating in the mind of the wholesale or retail buyer, the commission agent or the converter, those favourable first impressions of the goods that are so desirable, and in preventing the unfavorable impressions that must be produced by badly finished, badly inspected fabrics, made up untidily or irregularly, either in the piece or the bale.

In case a mill is sufficiently large to justify the expense, a cloth room should be located independently of the main building, preferably immediately between the weave room and the storehouse for baled goods and on the same level as these. It should be connected both with the weave room and with the storehouse by passages and should be of fireproof construction.

WEAVE ROOM

Where the weave room, the cloth room, and the passage between are all on one level, or approximately so, the cloth may be conveyed directly from the former room to the latter by means of trucks running on rails. When the weave room is situated above the cloth room, the cloth may be dropped into the cloth room by means of inclined chutes, while in case the cloth room is at a higher elevation than the weave room, some mills adopt an endless belt, usually of rubber, that travels at a slow speed and has lags attached at short intervals, to carry the cuts of cloth up to the cloth room. In the case of the storehouse, it is important that the floor should be of such a height above the ground as to afford facilities for loading the bales of goods into railroad cars or on to trucks without raising them, and for this reason it is not always advisable to have the storeroom floor on the same level as the cloth room; but there should not be sufficient difference in the heights of the floors to prevent the bales being trucked from the cloth room to the storeroom on an incline.

IMPORTANT FUNCTIONS OF A CLOTH ROOM

One of the most important functions of a cloth room is the inspection of all the cloth manufactured by the mill, so as to ascertain imperfections that necessitate the classification of any pieces as *seconds.* In order that the persons inspecting the cloth may perform their work to the best advantage, the light should be of the best and should come from the best possible direction. It is desirable that this light should come from the roof; a good type is a saw-tooth roof taking the light from the north side, or if this is impracticable, a monitor roof is desirable. In case it is impossible to have either, the cloth room should be so located as to enable the light to be taken from the north side, where the inspection tables should be placed. The artificial light supplied should also be ample and suitable for use when daylight is not available.

The cloth room should be suitably constructed to reduce the risk of fire to a minimum, should be provided with fire-prevention apparatus, be well ventilated, and suitably heated. As many of the machines required are of heavy construction and others have intermittent motions that produce considerable vibration, the floor of the room should be of heavy construction. There should also be below it a well ventilated space, such as a low story or a basement, to provide against dampness in the floor. If this is impossible, the cloth room should be provided with a sufficient number of slatted platforms on which cloth can be stored above the level of the floor. These platforms are divided by vertical slatted partitions into bins for the storage of different styles of cloth, after they have been folded and examined, until a sufficient number of cuts is collected to form a bale.

The arrangement of the machinery in the cloth room should be such as to reduce to a minimum the distances between the points where the cloth must be handled, always bearing in mind the necessity for adequate light at those tables or machines where inspection takes place.

Cloth rooms should be kept as clean as possible. If the condition of the room, as regards cleanliness, is not carefully considered and if dust and dirt are allowed to collect, it is evident that the objects of a large part of the work will be defeated by a simple matter that could be easily remedied by a little extra care and attention to details. For the purpose of keeping the cloth free from dust and dirt, the dust should be removed from the machines by means of a system of ducts and strong blowers or fans connected, to the different machines.

MATERIALS HANDLING

In order to reduce materials handling and speed the flow of work round a factory, various conveyor systems have been developed, ranging from relatively simple moving belts and manual push-pull rail systems to extremely sophisticated, computer-controlled automatic overhead conveyors. These are increasingly to be found in all areas of the factory, from cutting through sewing and finishing to warehousing and despatch. For example, one important area of current R&TD for materials handling in the cutting room is the robotic picking of cut pieces from the table and their placing onto automatic conveyor systems.

Fully computer contolled transport and picking systems are already well established in modern garment warehouses.

In garment assembly, the most modern Unit Production System (UPS) provide the capability to locate and track individual pieces and orders around a factory as well as re-allocate work in the light of changing resources, styles and work contents. However, they can be expensive to install and operate and there have been some spectacular and costly failures arising from attempts to introduce inappropriate systems. Overhead systems generally have large space requirements and operators do not like the physical and visual isolation which result from widely spaced work stations surrounded by 'walls' of garments. Supervisors also find that the line balancing and 'debottlenecking' capabilities of such systems are often far from ideal although this problem may reduce with the further development and integration of comprehensive CIM systems. A growing trend towards team work organisation of sewing and related operations may cause some rethinking of current directions in the area of materials handling.

FUSING

For certain classes of garment, especially men's tailored garments, the processes of fusing an interlining to selected cut parts, and final pressing of the fully assembled garment, are as at least as important as the sewing operation itself for the production of good quality articles.

Fusing and pressing are both processes which depend on heat, pressure and moisture, and the reaction of different textile fibres and structures to a wide range of varying conditions. In particular, new generation fabrics such as those based on microfibres with sophisticated surface finishes such as 'peach skin' effects are especially sensitive to both temperature and pressure. Extreme care has to be taken to prevent irreversible damage whilst ensuring a satisfactory and even bond strength. This is a complex field of textile technology which has not yet received sufficient attention. Most development to date has focused on the use of programmable micro-processors for better control, and workstation engineering to improve productivity.

One approach to improved regularity and versatility of fusing has been the development of flexible pneumatic roller systems which ensure completely even application of pressure across the whole fusing area, regardless of the thickness of the fabrics.

Parts of different thicknesses can even be processed at the same time, including pre-stitched components. Linked to this is the development of a much broader range of interlinings and adhesive systems. Interlinings are increasingly making use of the full range of new fibres and fabric forming techniques, including microfibres and specialised nonwovens. The precise selection and matching of interlinings to outer fabrics and to different applications and performance requirements is an obvious area for the application of objective testing methods and for use of expert systems.

SEWING

Various innovative methods of joining fabric and creating garment shapes have been researched in the past (*e.g.* radio frequency and ultrasonic welding, moulding and adhesives) but it is likely that, for the foreseeable future, clothing will continue to be assembled using sewing as the principal method of joining.

Nor is any major breakthrough in sewing technology likely. Developments here have tended to be incremental rather than fundamental and the basic sewing action seems likely to remain unchanged for the foreseeable future. These incremental developments fall broadly into two categories, in line with two main types of market requirement:

- Flexibility for quick response, with the ability to adapt to a succession of small orders in different styles involving changing stitch sequences and machine settings. To meet this need, a significant proportion of sewing machines are now pre-programmable, using microprocessors with push-button controls or, in special cases, responding to bar-code or "smart card" signals.
- Semi-automation for high productivity of a few basic or sub-assembly operations *e.g.* attaching pockets to jeans, stitching collars for shirts, hemming underwear and T-shirts. Most manufacturers have developed specially engineered workstations for very specific, basic sewing operations, often incorporating a degree of sensor technology and fabric manipulation, and enabling the operator to function more as a supervisor than a machinist.

The variety of competitive pressures and market opportunities ensures that both lines of development will continue. In particular:

- The facility to pre-programme individual sewing machines from a central computer will be incorporated into more ambitious systems of computer integrated manufacture (CIM). Ultimately, the capability will exist to programme machines with fabric-specific settings (based, for example, on Kawabata or FAST objective measurement data)
- Production engineering skills and developments in sensor and manipulative technology (often referred to as "pick-and-place" techniques) will lead to more obviously identifiable pockets of automation within the overall manufacturing sequence.

Sewing usually only accounts for some 10-30% of an operator's time. Handling and positioning of fabric pieces, which account for a much greater proportion of total time and effort, have therefore usually been the first areas to be tackled in the course of development of work aids and mechanisation.

The conventional basic flat sewing table which still accounts for the vast majority of all workstations throughout the world has always been a machine support rather than an ideal ergonomic configuration for handling fabrics. Workstations of the future may need to incorporate adjustable surfaces which can be rapidly 'programmed' to suit constantly changing tasks and garment styles and will need to be better integrated with conveyor systems.

PRESSING, FINISHING

Pressing remains a labour intensive 'craft' skill in many areas of the clothing industry and yet has a critical impact upon the final quality and value of the garment. Current technology developments are aimed at reducing processing times while improving performance and consistency of this operation. Ergonomic redesign of workstations by leading manufacturers is leading to the introduction of microprocessor controlled suction and central control units, variable height and table configuration mechanisms.

GARMENT DYEING

Garment dyeing techniques have been widely practised for a number of years. In Western Europe, Italy and the UK are leading producers, largely due to the influence of Benetton and Marks & Spencer and the importance of the knitwear industries of those countries. One estimate puts the total output of the garment dyeing sector at some 6% of all Western European apparel goods. The US market has traditionally lagged Europe in this field but is currently experiencing considerable growth.

Much of the recent boost in garment dyeing came from the fashion for a 'washed' look in denims and faded, creased cottons; before that, demand for fashion coloration of casual knitwear was one of the major driving forces of this technique. The garment dyeing sector is now contending with demands from other sectors of the apparel industry for rapid response service in higher quality and more technically demanding areas. There have recently been substantial improvements in the design and microprocessor control of machinery, fluid recirculation systems and the use of specialised dyestuffs and additives.

Coloration response times have been brought down from upwards of 10-12 weeks using the traditional fabric dyeing route to typically 2-3 weeks by garment dyeing.

However, there is still a considerable gulf between the response capabilities and economic scale of operation of even the most versatile garment dyers and the provision of individual colour choice on single articles at or near the point of sale. Very little work has so far gone into the development of microbath dyeing technologies capable of processing single garments in the minimum of dye liquor or other solvent. Patterning options at this stage are also very limited unless cross-dyeing techniques involving different fibres and dyestuffs can be employed. Qualitative developments are called for, not only in the technology of late coloration but in the whole organisation of materials sourcing, quality control and manufacture that precedes the garment dyeing operation.

Difficulties still experienced in the provision of 'custom' garment dyeing services include:

- lack of specifically designed machinery
- limited scope for colour matching and quality control at local level
- limited operator skills and experience
- garments not properly prepared after sewing and handling operations
- limited scope for subsequent easy care finishing
- environmental limitations on water and air emissions.

Where patterned designs are to be applied to a fabric or garment other than by colour weaving, knitting or cross-dyeing techniques, most conventional printing technologies are limited in their application to flat surfaces. The only exceptions are where quasi-2-dimensional sections of made-up garments can be presented to a printing head for application of localised block, stencil, screen or transfer prints.

All of these techniques require substantial economies of scale for the basic creation, engraving and colour formulation of each design and can rarely be adapted for truly individual coloration. Such methods are also usually limited to simple unstructured garments such as T-shirts, sweat shirts, hosiery etc. An ideal new technology would facilitate creation of coloured patterns directly on complex 3-dimensional shapes.

Cationisation techniques are already attracting significant attention from garment dyers. They originated with attempts to improve the dyeability of cotton using anionic reactive dyes to impart a positive, cationic charge to the fibre surface and proved so effective that, under appropriate conditions, untreated areas of fabric could be left substantially uncoloured or at least, show marked two-tone effects. This provides a basis for producing 'colourless' prints which can subsequently be 'developed' to any colour by a relatively conventional dyeing process. Treatment of loose fibre or individual yarns to produce mixed colour yarns and yarn-dyed effects is also feasible. Piece dyed denim can be produced by this route.

PROPORTION OF FINISHED GARMENTS

The proportion of finished garments is considerably greater for the U.S. 50-55% - than for the EEC -about 30% - or Japan with 20% or less. This selectivity on the part of the other developed nations has a great impact on both the trade balance and domestic employment. The result of these trading patterns was that, in 1983, world trade in textile products increased by only $700 million while U.S. imports grew by $1.3 billion. The past two to three years have seen the growth of a new import strategem, the use of uncontrolled fibers to circumvent quotas. Silk and ramie are the chief contributors and the sweater' market (about 600% imports) is the main target. International trade in textiles is governed by the Multifiber Arrangement (MFA), an instrument of GATT and the successor to the Short Term (STA) and Long Term (LTA) Arrangements Regarding Trade in Cotton Textiles. Under the MFA the U.S. has bilateral agreements with developing nations that spell out allowable import quotas and their growth.

The MFA was drawn up in 1972 with protocols revised in 1977 and 1981. The existing term of the Arrangement expires mid-1986. Its primary objectives are:

— To move sure the orderly and equitable development of trade while avoiding disruptive effects in individual markets.
— To give the developing countries with an increasing share of' world textile trade.
— To achieve the expansion of trade through the reduction of barriers to such trade.

These are worthy purposes and fully supported by the U.S. fiber/textile/apparel complex. However, as discussed above, if 1972 the U.S. textile industry was still in a fast growth phase, Taiwan and South Korea were still true developing nations, the. PRC was mainly shut off from world trade and Japan was seen as the principal threat. The apparel import penetration level was 18% and, for all textiles, 9% on a poundage basis. The suggested 6%/ year growth opportunity for developing country exports to the developed countries appeared reasonable. Neither of the Protocols renewing the MFA in 1977 and 1981 completely recognized the major changes that had taken place in the world textile economy.

These included not merely the growth of the major Far East exporters and their reluctance to open their markets to textile products, except on an as-required basis, but also their shutting out of the majority of developing countries whose exports to the developed countries-have displayed little if any growth over the past 10 years. If the PRC and the Big 3 exporters are excluded, U.S. imports from the remaining LDCs have averaged about 25% of all textile imports since 1973. Japan has averaged about the same percentage, while the EEC has averaged 40%. These factors all have contributed significantly to the inability of the present MFA to achieve its ends. A second major contributor is the failure of the bilateral agreement mechanism and its administration.

Imports from uncontrolled countries and of uncontrolled categories of goods from bilateral countries, transshipment, and fraud all have grown fastly in the past two years. The Reagan criteria of December 16, 1983 and the rapid increase in the number of calls have not clearly affected import activity. The rules of origin regulations will perhaps have a short term impact by closing one type of trade loophole—but others exist or will be developed.

COMPREHENSIVE APPROACH TO CONTROLLING

A more comprehensive approach to controlling the situation is explicitly needed if equitable and orderly growth is to be achieved and the viability of a basic, strategic industry is to be preserved through the use of trade agreements. No change can be expected in the pattern of U.S. demand for apparel in the foreseeable future when compared with the last 10 years. For an average GNP growth, population growth and constant dollar cash income (D.P.I. minus transfer payments minus other labour income, *e.g.* non-cash advantages) have shown good agreement with actual results over the past 20 years or more. Other models of imports and exports of garments and apparel-related products project imports and to a lesser degree exports to increase over time. The rate of enhancement in imports is dependent on the value of the U.S. dollar relative to other world currencies but also on U.S. economic growth.

Assuming a 3% per year increase in real GNP and a return of the strength of the dollar the 1972 level, imports are hopes to increase from 40% of the U.S. to 50% in the early 1990's. This projection could well prove conservative; other forecasts made by industry analysts range as high as 70%. Analysis of import trends displays that the dollar exchange rate elasticities for both garments and apparel textile products are decreasing. Historically, elasticities have averaged 0.9 for garments and 1.2 for textile products, whereas the current "point" elasticities are 0.8 for both categories. Thus the return of the dollar to more traditional values will not have as large a beneficial impact on imports as was formerly the case. The decline of both garments' and fabrics' sensitivity to exchange rate fluctuations may be attributed to the importers' learning curve.

Once the importer has setup and cultivated foreign contacts, both he and the supplier may be less sensitive to exchange rate fluctuations and more keen in maintaining a stable, long-term business relationship. It should be noted at this point that while models forecast 1983 imports accurately, imports in 1984 exceeded expectations by a significant margin. The over-stocked position of several retailers at the end of 1984 suggests that the level of imports in that year was an aberration and that more restraint will be exercised by importers in 1985 and 1986. Under almost any set of economic assumptions, however, the projected growth rate of imports is positive and this means a decline in domestic production of apparel. What, then, of non-economic factors? The forecasts discussed above all rest, implicitly, on the assumption that here after political actions or trade agreements will have the same impact as those in the past. However, the overindulgence of importers in 1984 triggered new levels of political and other defensive activities by the industry. Industry calls for a freeze or rollback of imports culminated in the introduction of the Textile and Apparel Trade Enforcement Act of 1985.

The major provision of the Act is that of limiting exports by the major exporters in 1985 to the levels that would have taken place if the provisions of the 1981 MFA had been enforced. In future years, growth of imports from the chief exporters would be limited to 1% per year and a control system developed to enforce this rate of increase. Other provisions comprised more generous access to U.S. markets for the smaller developing nations and extension of controls to comprise non-MFA fibers such as silk and ramie, which both had contributed importantly to import volumes in 1984.

Passage of the Act into law, even if changed, would have a tremendous impact on the domestic industry, allowing it to plan and finance its future with a confidence not now possible. Type greatest, and a very real, danger lies in the enforcement area. Should this be handled without determination, or if exceptions multiply, imports could very quickly become a problem again. Other political actions affecting imports were the stiffer Country of origin regulations announced in August 1984 and the country of origin labelling legislation of December 1984.

TRANSSHIPMENTS OF GARMENTS AND GARMENT PARTS

The former was necessitated by the increase in transshipments of garments and garment parts by off-shore suppliers to avoid import quotas. The latter was an integral part of the Crafted With Pride Council program designed to encourage the preferential purchase of U.S. made apparel. As has already been discussed, the current MFA and associated bilateral agreements have failed to prevent disruption of U.S. industry by imports. The MFA expires mid-1986 and the most occurrence is that it will be renewed, even though the GATT and several participating nations, which see it as a do stacle to trade liberalization, would prefer it to lapse. It is unlikely that the trading guidelines spelled out in the MFA will be changed greatly - the EEC seems satisfied with the status quo and the emphasis on liberalization by the LDCs will likely off-set any moves by the U.S. and Canada to toughen the restraint mechanisms.

The impact of the Trade Enforcement Act on the MFA renewal is difficult to estimate at this time, though its passage would have a pronounced effect on the bilateral process. Without such a change, the process and its interpretation by government agencies offers little hope for a meaningful change in the level or rate of enhancement of imports.

Problems of Import

One aspect of the Crafted With Pride (CWP) campaign has already been discussed, that of creating a greater demand for U.S.-made garments by the consumer at the point of purchase. This basic component rests on the belief that such demand is essential to preserving the two million Jobs, provided by the industry. A second aspect is to publicize the fact that, for several garment categories, importing is far costlier than is usually perceived and, when all the less obvious costs are taken into account, there is little to choose between the profitabilities at retail of imports and domestic goods.

The remaining components of the CWP program are more directly the relation of this book and, it is believed, can place the. U.S. soft goods industry in the strongest feasible competitive position. They centre on commercially available technology and are best introduced by outlining the present retail scene. A major component of apparel retailing strategy has long been to strive for market share by adding floor space. Properly stocked with the right merchandise at the right prices, success should follow. Conventionally, department stores held their dominant position with the help of well promoted branded goods, price points were well maintained and periodic sales events were well controlled and designed to move extra inventories. The chains controlled the lower price point, volume business but over time gradually moved up-scale, thus increasing. In recent past, the acquisition of national brands and strong promotion of house brands or private label goods by the chains cut further into the department store share and profitability.

In the interest of both exclusivity of merchandise and bigger mark-ons department stores relied more heavily on offshore sourcing and the chains followed. for this strategy, the discounters followed suit and at the same time intensified competitive pressures by themselves moving up-scale. By 1984, 30% of all garments sold were being sourced off-shore by the retailers themselves, by manufacturers, and by importing companies.

As exclusivity of design at a realistic price proved harder to find in the Far East, Europe was shopped more heavily and there the strong dollar was an added inducement. Howeyer, a high level of imports presents its own problems. These, comprise no returns, early payment and long lead times which increase the chance of stock-outs or markdowns. The growing significant of fashion with its attendant unpredictability adds to the problems.

The Fall/Holiday 1984 season which ended with an overinventoned retail trade and heavy markdowns has led many retailers to work on new business strategies, which recognize little or no future growth in apparel consumption and large industry surplus of floor space, but which must just with an increasingly style and value conscious consumer. The second group of major importers is the manufacturing industry which, without investing more than a token amount of capital, has been able to concentrate on balancing domestic against off-shore manufacture.

LABOUR CONTENT GARMENTS

At home, costs could be competitive for long runs on low labour content garments with better response times, which are attractive to retailers. For short run, labour intensive, or style-rounding lines, imports were the answer. However, here again lead times were a problem, as cutters not only had, to carry inventory but also were subject to returns from retailers. Kurt Salmon Associates has estimated that the cost to the whole soft goods pipeline of long work in-process times and slow supplier response times is of the order of 25 billion dollars per year. This figure comprises markdowns and discounts from fiber through retail as well as waste, mistakes, etc. Almost 3/4 of this loss is at retail and the biggest part comes from markdowns. It represents a huge profit opportunity for the industry if the 4-6 months fibers generally take to pass through the pipeline to retail sale can be cut even to 4-6 weeks. It is also a massive deterrent to imports, where the purchasing process begins 12-15 months ahead of the selling season. Quick Response (QR) technology is now available to the manufacturer, including CAD-CAM designing, laser cutting, computerized production routing and scheduling, unit production systems, and a large variety of specialized sewing aids.

Manufacturers who have installed such QR technology have seen up to ten-fold decreases in elapsed time from fabric received to garments shipped. Besides productivity gains, reductions in manufacturing cost of up to 20% are feasible depending on garment type and complexity. The principal deterrent to installing such systems for the majority of apparel manufacturers is the capital involved. Several are of the view that significant capital is unobtainable, the pay-back risks are too great, or they do not see themselves as asset managers. Others see their role primarily as marketers rather than as manufacturers. The driving force for increased use of technology by the manufacturer will perhaps come from retailers. As yet, they are unaware of the value of QR to the operations in a quantitative sense; yet the benefit improvement coming directly from the ability to place orders for goods a week before they are needed, for instance, should not be difficult to estimate. Short lead times not merely decrease inventories, but they also subside stock-outs and markdowns. Besides, they put imports by retailers in a totally new light, drastically reducing or eliminating any cost or service benefit.

The retailer who recognizes the opportunities QR offers can be expected to encourage the concept forcefully with his suppliers as he, in effect, passes many of his inventory problems to his supplier. The manufacturer who embarks on this technology path will not be slow to realize at least three things. The first is a reduced incentive to invest off-shore. The second is that importing is less attractive both for cost reasons and as he is now carrying a larger share of the inventory burden. The third and critical perception is that he must pass on his inventory problems to his supplier the mill or the converter.

Such a step will call for QR or Just in Time (JIT) responses by the textile producer and, in turn, by the fiber producer. The upstream members of the pipeline have invested very large amounts of capital in low cost, high productivity equipment over the past 10-15 years, when flexibility for manufacture of short runs was not a prime consideration.

Successful operation in a QR/JIT environment will call for a comprehensive review of alternative ways of doing business and will raise questions, not just of technology but of inventory management and other. disciplines. For instance, if the pipeline is to be shortened to the minimum, little or no time can be spent checking a supplier's material for quality. Thus, defining specifications and meeting them will become increasingly potent functional objectives.

So, too, will be an extended flow of information not only about a customer's schedules but also about those of his customers. It should also cover processing conditions at the operative's own and earlier processes, together with machine settings and conditions. The last aspect of operative control is related to the establishment of check procedures to see whether operatives are carrying out their quality duties according to the proper standards. The significant thing is that these checks should be systematic and provide results in such a form that action can be taken. This framework covers the chief areas that merit attention if quality control is to be effective. There will be marked differences of emphasis, depending on the section of the textile industry under consideration, and differences will also arise from firm to firm within a particular section.

TECHNIQUES EMPLOYMENT TO MAKE CONTROL

The framework within which techniques or procedures could be employed to achieve control of the quality characteristics of the product being manufactured. The range of subjects covered was considerably wider than that generally associated with quality control but was consistent with the wider definition of the subject given earlier. It is now necessary to consider the organizational implications of such a definition.

Management is liable for laying down an integrated policy with respect to quality improvement and control. This policy should cover all the widely varying matters. Above all, it should be consistent with the production policy, and both should be regarded as equally important. The danger is that a an explicit quality policy is not produced and the quality programme is left to the various departments and sections, with their conflicting interests. It should be admired that production and quality requirements are often initially inconsistent, and, unless management takes its quality liabilities as seriously as its production one, quality will go by default. To label the head of the testing laboratory or inspection department a 'quality controller' often aggravates the difficulty, as it can mislead management into believing that, once it has made such an appointment or has set up such a department, it has discharged its quality responsibilities. Almost every functional specialist has a part to play in assisting management to achieve the establishment and control of quality at the desired level.

In the absence of a clear policy, these specialists and their departments tend to work independently of each other. Even where no actual differences arise, decisions are taken by one section that affect the success of decisions already made in others. There are especial dangers in the situation where each functional specialist, such as the work-study engineer, the training officer, etc., believes that he has no direct responsibility with respect to quality, Since this is the liability of another specialist section—the quality-control department. The laboratory staff are often well aware of those factors that make control difficult, but, as they are usually of a similar status to the personnel from the other sections, it is obviously difficult for them to take the initiative for the necessary co-ordination. To avoid these difficulties, management should lay down a clear quality policy, together with the responsibilities of each section and department for the execution of such a policy.

What is needed is somebody able to give specialist advice on all aspects of control procedures relating to products, operatives, machines, work organization, etc. A feasible title, and one found increasingly in the United States and on the continent, is that of quality analyst. The word 'controller' has been deliberately discarded, since the real control must remain in the hands of the management.

The quality analyst covers a position corresponding to that of a production engineer on the production side. Practically, it may not be possible or economic to add a further specialist of this type, and, if this is so, it will be necessary to increase the authority and broaden the activities of the testing department. Top management would certainly need to call upon the quality analyst to assist it in evaluating and assessing the overall quality position, and it is therefore desirable to make him directly liable to the board of directors or the general manager. Once the policy has been agreed upon, it should be implemented in the same way as the production policy associated with it.

Control Quality

Quality control, as already defined, is concerned with both the procedures used and the action taken, to ensure that the quality characteristics of the product conform to the required standard. Consideration has already been given to standards, and the section on the framework of control has indicated the areas where control is essential. Following this, the organizational aspects of control have been discussed. Lastly, attention should be given to the actual procedures or techniques that the quality analyst and his staff can employ to assist management in taking corrective action.

Techniques Impact

The chief problem in describing such techniques is their number and variation, which represent, as they must, the requirements of widely different sections and processes. For example, techniques employed in the spinning section of the industry are those related to a process industry, whereas those in weaving are more analogous to those found in an assembly industry. Within each of these groups alone, there are a great number of procedures and checks.

No useful objective would be served by providing a catalogue of all the methods that could be employed in such a varied industry. It is more valuable to effort to uncover the features common to effective control. The first point to note is that there are very few so-called control procedures that do, in fact, control. With the exception of automatic-regulating devices that are built into the process, the effect of control procedures is indirect.

They serve generally to detect differences between actual and standard, and, for certain quality characteristics, to measure such differences quantitatively. Although this is an inevitable condition for control, it is clearly not a sufficient one. Control depends on the taking of the correct remedial action. It follows that a procedure cannot be judged merely from the standpoint of whether or not it reveals a special quality shortcoming; reference needs to be made to the likelihood that effective action will result from such a presentation.

Control Procedures

The first step in all control procedures involves the collection of data about the characteristics of the yarn or cloth being considered. It is at this initial stage that a serious problern is encountered. To obtain accurate data often involves the use or testing instruments of increasing complexity. Not only are the instruments complicated, but many can be used only under particularly controlled atmospheric conditions and hence must be placed in a room separated frorn the manufacturing process. It follows from this that the person who is carrying out the check is often different from the one who carries out the action demanded as a result of the check.

This extra step tends to militate against action. It raises an immediate communication problem, aggravated by a physical and, in particular cases, a status separation. The more complex the test procedures and the more specialized the interpretation of the results, the more difficult the problems become.

CLASSIFICATION OF CONTROL

Several points that follow from this directly affect the type of control procedures used. First, the best way to solve such a communication problem is to remove it by devising methods that can be employed by the people who are liable for taking action on the results. For instance, yarn cleanliness is often an important quality characteristic.

There are many sophisticated procedures in use in laboratories for judging this. It is difficult, however, to feed the assessment back to the mill in such a way as to promote remedial action. It is more satisfactory to devise a means of enabling the supervisors to check the amount of cleaning that is being performed by operatives; so that the essential action can be taken to remove the causes of dirty yarn.

There are difficulties involved in drawing up such procedures. First, it is difficult to assess cleanliness, especially when the degree hoped varies with both the fibre properties and the part of the machine inspected. Secondly, since cleaning is done periodically, to judge the operative objectively it is essential to inspect the machines over a long period of time.

Thirdly, the criteria of the assessment have to be sufficiently simple to allow the operative to appreciate in what way he or she is at fault and sufficiently quick to prevent the system from becoming to much of a burden for the supervisor. The method below shows how these difficulties can be overcome, but of greater significance is the fact that it provides a demonstration of the type at procedure that sometimes has to be utilised to instigate action. Initial agreement is reached on what are the key points needing inspection and, depending on the time available, on the number of machines to be inspected per day or per shift. A simple scale of assessment may be found to be most useful. From analysis, it is to see not only which operatives clean less frequently or less effectively or both than the others, but also which key points are being overlooked. It is also to judge whether work-study allowances for cleaning are for the degree of cleanliness required.

The supervisor knows where to take action and the operative can understand on what basis it is being taken. In several cases, it is not possible to use test methods that can be performed directly by supervisors. In these circumstances, it should be appreciated that whether action is taken or not on the ground of the results produced by the testing department depends on the confidence and understanding of the results by line management. Every effort must be made to develop and use simple techniques. An illustration of this is found in the example of machine control below. Simple machine-control methods are of great value to supervisors, especially in an industry where they are often liable for a large number of machines. There are various statistical methods for revealing machine differences, but these are often too complicated for departmental supervisors to understand.

4

Spinning Yarn Fabric Construction

When bundles of fibres are drawing out and twisting by a process called spinning. In the starting, the yarns were spun by man with bare hands without the help of any tools and it must have been several centuries before the spindle was evolved for spinning.

The spindle or 'Takli' still survives for spinning wool, silk and cotton yarn. It is the simplest tool used for spinning which comprises a round disc which is attached in the centre to a thin, smooth rod about 7 inches long. The upper end of the rod has a groove or a hook which catches up the fibres. When the spinner pulls the bulk of the fibres, these are drawn out in a long strand. The spinner while simultaneously pulling the fibres, provides a twist to the spindle and lets it go. The spindle whirls round, and thus, twisting up the pulled fibres makes a continuous thread.

The thread is then woven round the rod above the disc. The disc gives the essential weight which quickens, increases and prolongs the revolving of the rod, thereby, the twisting of the fibres ensures a strong thread.

Spinning Wheel or Charkha

Who does not know the, *'Charkha'* in India? A *Charkha* is for spinning yarn in the handloom industry in India. The yarn spun or the charkha is of different qualities. The famous 'Dacca muslins', which were uncomparable for their fineness, but alas are extinct now, were woven of a charkha yarn. The immitable soft and light pashmina of Kashmir are still woven of the charkha spin yarn.

In the textile industries today, electrically driven machines are employed for spinning. Many machines are used to complete the process of spinning, which of stages, such as drawing, out of the fibres in silvers, further drawing out of silvers, further reduce size and to give slight twist-roving and then spinning.

There are two methods of procedure. In one process, the action of drawing, twisting, and winding is regular and this is known as 'ring' spinning, and in the other, the draw and twisting is stopped while the twisted thread is wound up (as in the case of hand spinning) and this is known as mule spinning'.

The 'ring' spinning is a quicker process and has the benefit of reducing operating cost and increase production, but the mule spun yarn is finer, softer and of greater evenness. In spinning yarns, whether hand or machine, a difference has to be made in the yarn intended for warp and weft or filling.

The wrap yarn requires a greater amount of twists to produce strong, firm thread which is used for the foundation of the fabric and is subjected to mire strain and friction during the process of wearing.

The yarn is provided twists of a particular number of turns to the inch and are given either 5 or 2 twists. At first, regular thread or a strand of twisted fibres is made and save ral of these strands are twisted together to get the final yarn.

Count

The yarn is described forsize or count. A skein of yarn, 840 yards in length, is known as a hank and is the ground for determining the count. If a hank of 840 yards weighs one pound, the yarn is number one, and if two hanks weigh one pound the yarn is of count 2. Several variations are introduced in the preparation of yarn by mixing different sorts of fibres, by mixing various coloured threads or to produce texture, decoration or to add particular properties to the fabrics. The variations in yarn may be divided into two main groups:

- Novelty yarns.
- Simple.

Novelty Yarns

The construction of these yarns is of a complex nature and is varied in many ways. These yarns are usually ply yarns of different kinds of fibres or of separate colours and are irregular rather than smooth.

Single strand or yarn of various colours, sizes or fibres may be twisted together to form one complex yarn. Another kind is brought about in this type of yarn by varying the tension or speed after, intervals of certain length thus, allowing one part to loop or twist around the other. Novelty yarns are also prepared from simple yarn by varying the amount of twist. The complex type of novelty yarn is used with two objects in view, one is to combine different fibres *e.g.,* cotton and rayon may be blended with or covered by wool or silk.

This lowers the cost of production. The other objective is to produce a novelty yarn for the construction of novelty yarns, at least one or two single yarns are used, one forms the foundation yarn known as a *base* or the *core* and the other, the effect yarn, which is wound or looped round the first one. A third yarn knownas binder yarn is often used to fasten or tie the effect yarn to the foundation yarn. These types of yarns are mostly used for drapery and unpholstery fabrics.

Simple Yarn

In the construction of simple yarn, only one type of fibre is used. The manner in which the fibres are twisted will be the same throughout the length of the yarn. Yarns, called simple, ply or cable, depending upon the number of strands they contain.

Ply Yarn

Two or more than two yarns are twisted together to form a ply yarn. These yarns are called multiple strand yarn. If two single yarns are twisted together, the resulting yarn is called two-ply yarn, if three are twisted together, three-ply yarn and so on.

Single Strand Yarn

In this, so many of fibres are twisted together into a continuous length. The yarns consist of one kind of fibre and of one colour. This kind of yarn is the one usually found in most standard fabrics for clothing and household use.

Kinds of Yarn

Knot or Spot Yarn

In this also two strands or yarns are used. One is the base, the other is utilised for forming knots or spots by an additional turn round the base yarn.

Slub Yarn

This yarn is composed of soft untwisted places at regular intervals. This may be a single strand yarn or may comprise two or more strands

Loop Yarn

This yarn comprise loops at intervals round a coarse foundation.

Corkscrew Yarn

This yarn is constructed by twisting together yarns of various diameters or by varying the rate of speed or the direction of the twist.

YARNS MANUFACTURING

The twist of yarns and threads influences their morphology. This feature creates the thread's properties, and at the same time is decisive regarding the processing throughput. A great amount of the literature published hitherto deals with the problem of yarn twisting, but there is a lack of sufficient information related to the twist of multi-folded threads. Technologists in the textile industry set the thread's twist value in their 'spinning planes'. Its actual realisation on the spinning machine and ringtwisting frame is performed by selecting the spindle rotary speed and fibre stream delivery velocity. Yarn plying, i.e. twisting together some component fibre streams, causes a decrease in the linear mass irregularity (thanks to a better fibre packing in the stream), an increase in tenacity, a lowering of the bending stiffness, an increase in the abrasion resistance, and a lowering of the thread's tendency to pilling formation. Owing to these favourable circumstances, twisting positively influences the thread's manufacturing throughput and the barrier ability (e.g. against UV radiation) of the fabrics produced from such a yarn.The aim of our work is to compare the properties of various yarns and multi-folded threads manufactured by means of new spinning technologies, and to analyse their twisting processes.

ANALYSIS OF THE TWISTING PROCESS OF MULTI-FOLDED YARNS AND THREADS

The investigations so far carried out in Poland and abroad have focused mainly on the analysis of the process of yarn twisting. The research results have been presented in a monograph. The theoretical analysis of multi-folded threads has come within the investigation scope of only a few researchers, and its considerations are included in the following works. On the basis of an analysis of these considerations, we can state that the twist of the component fibre streams which form the thread depends on their radius Rs, the assumed (set) twist value Sp of the multi-folded thread, and the coefficient °, which characterise the material parameters of the component fibre streams. The relation between these parameters is described by the following equation.

The twist level of multi-folded yarns and threads depends on the properties of the raw material used for their production, including its linear mass, the fibre stiffness, and the structural features of the component fibre streams. The chief point of the problem analysed concerns the selection of appropriate twist values: Sp of the multi-folded threads, and Ss of the component fibre streams, which will be decisive in relation to their technological usability, described by the engineering denotation of 'processing throughput'.

It is accepted that the twist value Ss of the component fibre streams is 20% lower than the twist value of the multi-folded thread [3, in 2003]. According to: ±ms = 0.55±mp, and Ss = 0.8Sp. Twist is also the parameter which can aid the achievement of better protective features against harmful UV radiation of the final product. The selection of an appropriate plying direction is an essential problem connected with twisting the component fibre streams with the use of ring-twisting frames.

The twist direction of the multi-folded thread can be in accordance, or not, with the twist direction of the component fibre streams. The effect of using twist on twist (S/S, Z/Z) is an increase in the twist values of the component fibre streams below their convergence point. The multi-folded thread shaped in this way is characterised by special stiffness and great ability to form loops. In contrary, the effect of applying opposite twists (Z/S, S/Z) is the complete or partial elimination of twists of the component fibre streams below their convergence point.

The effect of this phenomenon is that the component fibre streams tightly adjoin to each other, and form almost one component fibre stream. The multi-folded thread shaped in this way is softer, has a higher breaking tenacity, and is characterised by a lower ability to form loops. The users' increasing demands in relation to textile products and the changing fashion trends are forcing yarn manufacturers to permanently improve the yarn quality and broaden the range of assortments. At present, ring-spinning frames equipped with modern technical solutions of the working elements are available on the market. These spinning machines enable the manufacture of new types of yarn and multi-folded threads.

METHODS OF YARN FORMATION

Yarn formation methods were originally developed for spinning of natural fibres including cotton, linen, wool and silk. Since the overall physical characteristics of the fibres and processing factors needed differed from fibre to fibre, separate processing systems were developed. As synthetic fibres were introduced, synthetic spinning systems for texturised and untexturised cut staple were developed as modifications of existing staple systems, whereas spinning systems for texturised and untexturised filament were developed separately.

Staple yarn formation involves multiple steps and can include:

- Fibre cleaning and opening (as needed for natural fibres);
- Fibre blending (to assure uniform mixing in natural fibres or in fibre blends);
- Carding (to align fibres and to remove short fibres);
- Combing (if highly aligned fibres are desired);
- Drawing and spinning (to reduced the denier of the yarn, to provide twist and to give cohesion to the yarn); and
- Doubling or plying and twisting of the yarns (as needed to provide greater uniformity).

In recent years a number of staple spinning processes other than ring spinning have been developed that reduce or shorten the number of steps necessary for formation of yarns suitable for textile substrate formation and are discussed separately following conventional ring spinning techniques. Yarn preparation from fibre filaments is much less complex and often no or only limited twist is imparted prior to use in the textile substrate. The steps involved in yarn formation are outlined.

COTTON BALES SYSTEM

When cotton bales arrive at the textile mill, they are highly compressed and have been ginned to remove seeds and some of the impurities present. On removing the cotton from the bales, the cotton is opened up, blended and mixed with cotton from other bales. During the opening and blending process, the fibres are separated and loosened from each other, trash is removed from the fibres, and the fibres are more randomly mixed to assure greater uniformity. Finally the fibres are formed into a thin partially oriented continuous web of intertwined fibres called a picker lap.

The picker lap in turn undergoes carding to remove short fibres and remaining trash and to provide additional orientation to the fibres. Carding involves pulling, separating, and orienting the fibre s by passing the lap between successive cylinders moving at different speeds and containing fine bent wire bristles that catch the fibres.

The carded lap is removed by a doffer cylinder in the form of sliver (a rope-like fibre mass) and coiled into a rotating can. The sliver may undergo additional blending to improve its uniformity and density so that it may be more effectively drawn and spun into yarn.

At this point, the sliver is suitable for drawing and spinning into medium and coarse yarns. Before drawing and spinning into fine yarns, the cotton sliver must first be combed to further straighten and orient the fibres and to remove additional short tangled fibres. The drawing and spinning process involves passing the sliver between and through a series of rollers moving at progressively higher speeds to draw the sliver to a finer, more oriented, and uniform structure followed by twisting as the sliver is played onto a turning spindle. The drawing portion of the operation is referred to as the drafting process.

The degree of twist will depend on the speed of the turning spindle with each complete turn of the spindle providing a single complete twist in the yarn. In the initial stages where little twist is present in the drawn sliver, the sliver is fed through a tube and onto the spindle. This process is called roving. Subsequent drawing and high speed twisting is carried out by ring spinning in which the drafted, lightly twisted sliver (roving) is fed from the drafting unit onto a high speed spindle via a traveller holding the spun yarn to a ring surrounding the reciprocating spindle. The traveller can move easily around the ring and provides a slight drag on the yarn as it is fed onto the spindle. Ring spinning proceeds at 5,000 to 10,000 revolutions per minute.

THEIR YARN-FORMING SYSTEMS

Since conventional ring spinning systems take several steps to reduce fibres to a spun yarn, several alternative techniques have been developed that reduce the number of these steps, Most of these processes have more limited usefulness with regard to the range of linear densities and type and distribution of fibre lengths that can be processed by these systems,

Open-End or Break Spinning

Open-end or break spinning systems have come into wide use for spinning of short and medium staple fibres directly from sliver to yarn in one step without carding or roving. Sliver is fed into a breaking or opening unit to separate the fibres from one another, and the fibres are then forced by air pressure into a hollow rotor rotating at high speeds (up to 50,000 rpm). The fibres are deposited by centrifugal force on the sides of the hollow rotor, and the resulting yarn is removed continuously by a stationary tube mounted with in the rotor.

The rotating rotor provides twist to the yarn and produces a yarn with somewhat greater higher pitch and bulk and somewhat lower strength than ring spun yarns. Open-end spinning operates at a rate up to five times that of ring spinning and can be effectively used for cotton, polyester-cotton blends, as well as other short and medium staple systems. Synthetic staple fibres such as polyester alone can not be effectively open end spun due to dusting of oligomer from the fibres that interferes with the spinning action of the rotor.

Friction Spinning

Friction spinning is a variation of the open-end spinning system and often referred to as the DREF system. In this system, one or more slivers are fed onto a rapidly rotating card drum which opens the sliver to form single fibres. The separated single fibres are blown from the card drum by a stream of air onto the junction of two parallel perforated drums turning in the same direction. The rotating perforated drums under suction cause the fibres to be compressed and twist around one another to form a uniform yarn which is continuously removed onto a spool. The resultant yarns are bulky and have properties similar to woollen yarns.

Air-Vortex Spinning

Air-vortex spinning is similar to open-end spinning but utilises a stationary tube rather than a rotor. A high speed air vortex is created in the tube that deposits fibres with in the tube and simultaneously provides twist due to the vortex. The yarn is continuously withdrawn as in the case of open end spinning.

Fasciated Spinning

In fasciated spinning long staple sliver without twist is introduced into a limited space and subjected to a torque jet operating at right angles to the flow of the sliver thereby imparting a false twist to the sliver. As the sliver exits the torque jet it rapidly untwists, and the outer fibres tend to break away from the sliver and wrap around the inner sliver to give a strong yarn consisting of mostly parallel fibres with some fibres tightly twisted around the outside.

Self-Twist Spinning

In self-twist spinning, two parallel slivers are fed between two reciprocating rollers which form identical left hand or right hand twists in each of the slivers alternating down the length of the sliver. The resulting false-twisted yarns are then brought together so that the right hand twist segment of one sliver is phased with the left hand twist of the other sliver. On relaxation, the slivers untwist over one another to form a stable yarn. The process was originally developed for spinning wool, but has been used extensively for acrylic yarns. In the related Selfil spinning method the self-twist yarn is wrapped by alternating phases of continuous filaments to form a highly stable wrapped yarn containing less than 10 per cent filament.

Covers Pun Spinning

Covers pun is both the name of a spinning system and a yarn. In this system staple rovings are drafted in a conventional manner, and then the roving is passed into a hollow vertical spindle.

On the outside of the spindle a filament yarn, from a cylindrical spool rotating at 20,000 to 30,000 rpm, is fed into the top of the hollow vertical spindle with the roving. The rotating filament spindle causes the filament to wrap around the core of the staple to produce a wrapped yarn consisting of 80 per cent to 95 per cent staple. Polyester filaments are usually used to wrap pure yarns or blends of cotton, wool, nylon, and acrylic staple to form sewing threads or yarns for textile substrate production.

Integrated Composite Yarn Spinning

A composite yarn is formed by melt extruding fibres from a spinneret or by coating filaments with a molten polymer followed by coating the emerging fibres with short staple fibres. The resulting matrix is immediately twisted causing the staple fibres to imbed in the extruded fibres before cooling. The resulting composite yarn is formed at a very rapid rate on the Bobtex spinning apparatus developed for this process.

Twistless Systems

In twistless systems liquid or powdered polymer adhesives are applied to sliver or filament tow and the adhesive activated by heating or steam to cause the individual fibres to adhere to one another. In some systems, after textile substrate formation from the twistless yarns the adhesive is removed to improve the aesthetics of the resulting textile. Adhesives used include polyvinyl acetate, polyvinyl alcohol and starch.

WOOLLEN AND WORSTED SYSTEMS

Wool is spun into yarns by either the woollen or worsted system. The woollen spinning system is less complex than the worsted system and utilises shorter wools of a wider range of lengths and diameters to give a low twist bulky yarn. The worsted system produces highly twisted fine yarns utilising fine fibres of a narrow distribution of length and size. Fabrics made from woollen yarns tend to be bulky and contain more entrapped air; whereas fabrics from worsted yarns tend to be tightly woven and fine with a hard flat surface.

On arrival at the mill as bales, raw wool contains large amounts of grease, swint (salts from the body of the sheep), dirt, vegetable material and other impurities.

The wool must be washed in successive baths of detergent solution to remove these impurities. The process is called scouring, and the weight of the raw wool can be reduced by as much as 50 per cent by the scouring process. Wool grease (lanolin) is effectively recovered from the scouring liquor as a commercial product.

Vegetable matter remaining in the wool can be removed by passing the wool through concentrated sulfuric acid to chemically destroy the cellulosic matter, a process called carborisation. After washing and drying, the cleaned wool is blended and carded.

At this point the sliver can be drawn and slightly twisted to form a roving which can then be spun into a woollen yarn. The sliver must undergo additional straightening, orientation, and removal of short fibres to be used in the worsted system. This process involves several successive steps including gilling (a form of pin orientation) and combing to give wool top. The wool top is drawn and slightly twisted in several stages to form a roving which is finally spun into a highly twisted worsted yarn.

OTHER STAPLE SYSTEMS

Other natural staple and synthetic fibres can be spun on cotton and wool systems if these systems are modified to take into account unique factors such as fibre length, crimp and linear density. Cut staple manmade fibres arrive in boxes at the mill and are ready for carding and processing into yarn. When two or more different staple fibres are mixed, it is critical to provide extensive blending before carding and repeated doubling of the sliver to assure intimate blending prior to roving and spinning. Other spinning methods that take fewer steps have been developed for staple spinning and are discussed under other yarn forming methods.

FILAMENT SYSTEMS

Filament spinning systems are much less complex because the fibres are continuous and do not need to be highly twisted to give a cohesive strong yarn. Filament yarn spinning usually involves man-made fibres and only the portion of the ring spinning system that involves twisting and winding onto spindles is used.

YARNS MANUFACTURED WITH THE USE OF A COMPACT RING-SPINNING FRAME

Spinning with the use of compact ring-spinning frames is a modified version of spinning with the use of classical spinning machines. Compact ring-spinning frames are equipped with compact drafting arrangements [13, in 2003], which enables an improved process of fibre incorporation in the yarn core.

Yarn manufactured with the use of compact ring-spinning frames is characterised by lower hairiness compared with classical yarn. Thanks to the compaction of the fibre band in the compact drafting arrangement, the fibres are better packed in the yarn, which means that the yarn cross-section includes more fibres. The result is lower linear mass irregularity of the yarn. The yarn is more compact, has better tenacity and higher elongation at break. Better structure perfection, better surface smoothness, and a smaller number of protruding fibres cause better properties of the compact yarn compared with classical yarn. At present, modern spinning machines are manufactured which enable the manufacture of compact core yarns. Such yarns achieve higher tenacity by applying a core consisting of filaments.

MULTI-FOLDED THREADS

Twisting a multi-folded thread composed from two component fibre streams; the directions of the tensions T0 and T, and the torque momenta Q0 and Q, which all affect the streams.

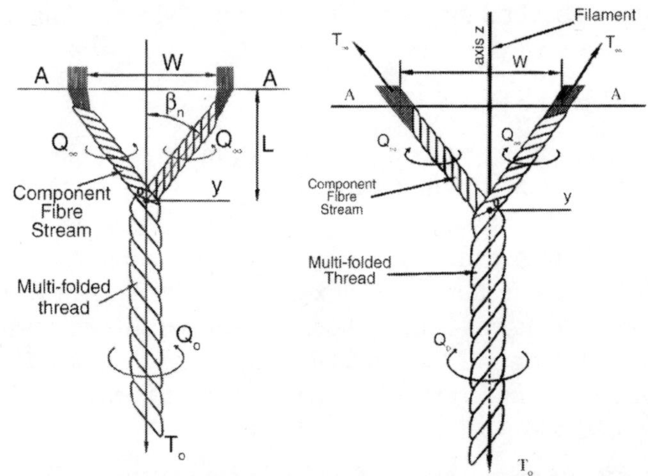

Fig. Scheme of a twisted classic multi-folded thread **Fig.** Scheme of a compactmulti-folded thread with core

The tension and the torque momentum of the multi-folded thread is described by the following equations:

$$T_0 = 2T_\infty \cos\beta_n + 2(K - B)\cos\beta_n \sin^4\beta_n - \frac{1}{2}\cos\beta_n \sin^2\beta_n$$

$$Q_0 = 2T_\infty \cos\beta_n + 2K\cos^4\beta_n \sin\beta_n + 2B\sin^3\beta_n\left(1+\cos^2\beta_n\right) + \frac{1}{2}\cos^2\beta_n \sin\beta_n$$

$$Q_0 = T_0 \tan\beta_n + 2K\sin\beta_n \cos 2\beta_n + 4B\sin^2\beta_n + \frac{1}{2}\sin\beta_n$$

The dimensions of the twisting triangle depend on the distance of the convergence point O from the line A-A of the delivery rollers' grip and the distance W between the two component fibre streams. When the convergence point is near the line of the delivery roller grip of the drafting arrangement, the level of the spinning tension is low (the distance W between the component fibre streams is small). Only very small dimensions of the twisting triangle allow small breakage of the multi-folded thread to be achieved. To achieve higher quality requirements, for example, multi-folded threads are manufactured from compact yarns twisted together. Spinning machines which manufacture compact multi-folded threads are aimed especially at wool spinning, considering the relative low breaking tenacity ??of wool. During condensing, which takes place in the compact drafting arrangement, both component fibre streams near each other and take on a V-shape.

Fig. View of yarn manufactured by different spinning systems.

The condensation of the component fibre streams takes place at a segment of two skew slots, under the action of negative pressure, by the turn of both streams around their axis. As a result, the phenomenon of fibre protruding from the component fibre streams does not take place, and nor does the twisting angle exist. With the aim of improving the quality of textile products, it is possible to manufacture compact multifolded core threads. Figure 2 presents a scheme for spinning a multi-folded thread composed of two component fibre streams and a core. In the figure the directions of the tensions T0 and T, and the torque momenta Q0 and Q"", acting on the fibre streams, are drawn. By applying a core composed of filaments, the multi-folded thread achieves better strength properties.

Yarn manufactured using the following spinning systems:

- Siro-spun ring spinning *frame yarn*
- Single compact yarn
- A multi-folded thread-compact yarn
- A multi-folded thread-twisted compact yarn
- A multi-folded thread-twisted classic yarn

A multi-folded thread according to the literature, is described as having the following qualities: very smooth structure, larger gloss, soft hand, high regularity, low hairiness (especially of long fibres – value S3 >3), possibility of spinning with lower twist factor ±mp, reduction of twist value Sp up to 20%,thus keeping strength of thread on the level of classic thread, as well as high abrasion resistance.

The above-mentioned considerations concern the component fibre streams obtained with the use of flyer-roving frames and ring spinning frames. A multi-folded thread can also be manufactured from yarns obtained from rotor spinning machines. However, it should be taken into account that these yarns are characterised by a different structure compared with yarns obtained from a classic ring-spinning frame, which results in their diverse properties [2, in 2003]. Thanks to the nature of the process of rotor spinning, the manufactured yarns are distinguished by a sheath, which means that they have fibres wrapped around their surface; this is the reason that rotor yarn should be manufactured with the assumption of a higher ±ms coefficient. Figure 4 schematically presents two rotor yarns which have the same ±ms coefficient and angle [2] of the fibre inclination in relation to the yarn axis, but which differ by linear mass, a) smaller, and b) greater.

According to analysis, for example, for ring-spun knitting yarns, the ±ms coefficient is accepted within the range from 60 to 90, while for rotor yarns within the range from 110 to 120. In the case of warp weaving ring-spun yarns, the ±ms coefficient is accepted within the range from 115 to 135, and for rotor yarns from 135 to 155. The use of rotor yarn as a component fibre stream in a multi-folded thread may allow us to lower its twist value, which in turn results in an increase in the spinning machine's efficiency. Spinning technologists aim to manufacture linear textile products which should meet specified criteria. One of these criteria may be to secure protecting barrier properties against UV radiation. This can be realised by using appropriate raw material in the process of ring spinning (the best thin fibres) and selecting the yarn twist value. Taking advantage of the qualities of multi-folded compact yarns and threads allows us to create a totally new product with new properties. The advantages which result from manufacturing and applying compact yarns and threads, compared with the classical ones, include a lower breakage number, a decrease in the average percentage of good fibres in waste, lower dustiness of the manufacturing rooms in spinning mills, weaving departments, and knitting plants, low pilling ability of the final products, and higher machine efficiency. In the weaving process, the warp is exposed to different, diversified cyclic destructive loads during its path over the loom. These include repeatedly occurring stretching and bending tensions, impacts, and friction forces. Considering these peculiarities of the weaving process, compact yarns and threads are especially suitable for the production of woven fabrics. The twist of yarns and threads influences the protective properties of textiles, and during the ringspinning process, it has a significant effect on the dynamic phenomena which occur in the twistingand- spinning unit of the spinning machine.

THE INFLUENCE OF COMPONENT YARNS' CHARACTERISTICS

A fancy yarn is a textile yarn with virtual-ly unlimited pattern designs. Fancy yarns present deliberate, decorative continuous or programmed effects of colour and/or form, and they are used to create some variation in the aesthetic appearance of a fabric or garment.

Generally, the yarn structure comprises a basic yarn, one or more effect yarns producing regularities or irregularities in the yarn structure, and a binder yarn to fix the basic and effect yarns. The structure, and therefore the appearance of the fancy yarns, may vary according to the process-ing technology, raw material, number of component yarns and machine settings.

Fig. Ring Twisting Frame

Fancy yarns can be produced on spinning systems and twisting systems. There are three different twisting methods:

- Ring fancy twisting
- Hollow spindle fancy twisting
- Combinations of the hollow spindle and ring twisting processes.

The PL 31C ring twisting frame has been used in this study to produce three sets of fancy yarns (spiral, loop and bunch yarns), according to arotal plan of ex-periment (5-leveled plan with á = 2), making a total number of 56 variants (15 spiral, 15 loop and 26 bunch). Each component yarn was 100% cotton, 20 tex. The yarns were differentiated in respect of the component yarns' characteristics and machine settings as follows: linear density, twist, the angle of lap of pegs by yarn on the slat of the ring twisting frame (the input only for spiral yarns), overfeed of components (the input only for loop yarns and bunch yarns) and the frequency of the rocker's action in the ring twisting frame (the input only for bunch yarns). The following parameters were measured as output parameters: the linear density of the fancy yarns, tensile properties, bending rigidity, air index, snarling ability and the distribution of effects along the yarn.

PROCESSING THE YARNS

Spiral Yarn

In a spiral yarn, one component binds the other (which is straighter), thus creating spirals along the whole length (Figure 1). A spiral yarn is created as a result of the multiple twisting of two components with relatively slight overfeeds of one of them, or two components with the same or a slightly different linear density, and a different twist direction.

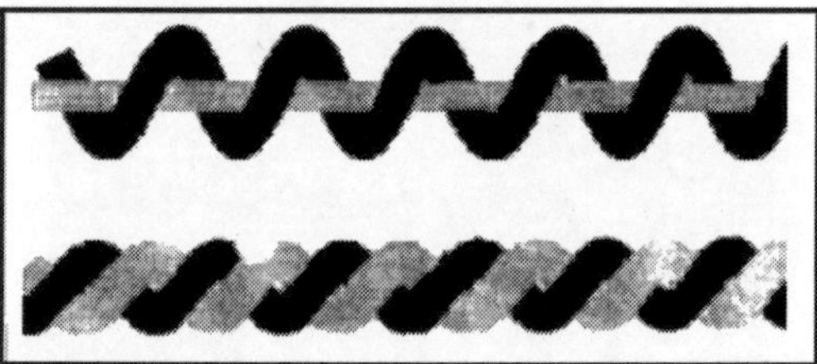

The spiral yarns produced for this study consist of two cores, white and black in colour, with a linear density 20 tex and twist direction Z. During the produc-tion process, the cores remained paral-lel before the twisting zone. From one to five effect yarns (20 tex each, twist direction Z) were added, according to the rotal plan of the experiments.

They had a different, contrasted colour (light yellow, dark yellow, orange, light blue and green) in order to identify them more easily when analysing the fancy yarns' structure. Both core and effect yarns were twisted in the S direction. The next important step was the binding process. A binding yarn (20 tex, in a violet colour) was twisted with the cores and effect yarns in the Z di-rection, each time with the same number of twists per metre, as in the process of twisting the cores with the effect yarns.

METHODS FOR METROLOGICAL ANALYSIS OF THE YARNS

The following parameters were taken as output: the linear density of the fancy yarns, tensile properties, bending rigidity, air index, snarling abilities and the distri-bution of effects along the yarns and twist. The following apparatus and classical or innovative methods were used to assess the following metrological parameters:

- Linear density of the fancy yarns: ITF maillemeter was used, according to ISO 7211/5 (1984).
- Elongation and breaking force of the yarn: Statimat M was used. 10 tests were performed for each yarn. The speed of the drawing frame head was 250 mm/min, and the length of the test specimen was 250 mm
- Bending rigidity: the Kawabata sys-tem was used.
- Air index: the Air Index Tester was used. This test checks the ability of the air jets to carry the yarns through the nozzle of an air-jet weaving ma-chine. Specimens of 30 metres' length were passed through the nozzle with 5 windings for spiral and loop yarn, and specimens of 2000 metres' length were passed through the nozzle with 10 windings for the bunch yarn.
- Snarling abilities: a snarling tester (no norms elaborated) was used. A speci-men of 50 cm was fixed between 2 clamps. The moving clamp was drawn in the stationary clamp's direction un-til snarls appeared.angle- represents the angle of lap of pegs by the yarn on the slat of the ring twisting frame; the pegs allow the components to be introduced into the twisting zone under the required angle (Figure 2).

Loop Yarn

Loop yarns have round, regular shape loops on their surface, of identical or different sizes. They are produced by the twisting process only with a big over-feed of one component yarn. Loop yarns consist of three component yarns: two core yarns and one effect yarn, twisted together with a binding yarn (Figure 3). Loop yarns were produced according to the plan of the experiments, having a lin-ear density of one component (tex), twist (twist/m) and overfeed (%) as input pa-rameters (Table 1). The production proc-ess was similar to that used in producing spiral yarns.

Bunch Yarn

A bunch yarn (Figure 4) is made of two component yarns twisted together, one of them (the core) being introduced into the twisting zone with a locally higher ten-sion. The effect yarn creates effect points, locally compacted, at regular or irregular distances. Such effect points may have different colours, shapes or sizes. The bunch yarns were produced, according to the plan of the experiments (Table 1) with the following input parameters: the linear density of the components (in tex), twist (in twist per metre), overfeed (in %) and rocker frequency actions per pro-duction cycle. The rocker decides the frequency of the effects points.

Statistical analysis of the results and discussion

All the metrological results of measure-ments obtained were analysed using the SPSS 12.0 for Windows program; multi-ple regression module statistical models were obtained. The following conclu-sions were found and discussed below for each type of yarn.

SPIRAL YARN

- There is a linear dependence with determination coefficient R^2 = 0.646 on a level of significance of 0.01 between the thickness of the spiral yarn, measured at the point where the binding yarn covers the cores and effect yarns and collapses into them, and the twist measured in the place on the yarn where no collapse oc-curs. This dependence suggests that the twist applied before and during the bind-ing process but in opposite directions has an influence on the final thickness measured at the point where the bind-ing yarn collapses in cores and effect yarns. The higher the value of that twist, the lower the thickness value. Thus, one may ascertain that giving a twist, first to the effect yarns and then to the binding yarn, causes compression at the point of contact between the binding yarn and the effect yarn, which in turn causes the convexity of the fragment without the binding yarn collapsing.
- There is a linear dependence with determination coefficient R^2 = 0.650 on level of significance 0.01 between the breaking force and the linear density of the spiral yarn. The higher the value of linear density, the higher the value of breaking force of the yarn.
- There is a linear dependence with a determination coefficient of R2 = 0.650 at a level of significance of 0.01 between the bending rigidity and the binding twist as well as the linear density. An increase in the linear density and the twist of the yarn causes an increase in bending rigidity. The dependence is linear. The linear density of the yarns has a higher influence on the bending rigidity than the twist, when the very restrictive level of significance of 0.01 is taken into account.
- There is a statistical dependence be-tween the linear density of the yarn and the width of hysteresis which appears in the bending rigidity test. Statistics show that the higher the linear density, the wider the hysteresis loop. The lin-ear density is an excellent descriptor of the width of the hysteresis loop. The dependence is linear. This

phenomenon is closely related to the bending rigidity, as mentioned above in point 3. The twist does not influence the widening of the hysteresis loop when the very restrictive level of significance of 0.01 is taken into account.

- There is a linear dependence (determi-nation coefficient $R^2 = 0.951$ at a level of significance of 0.01, standardised coeffi-cient Beta = 0.981) between the standard deviation from the average distance between the clamps of the device used for the snarling test, at the moment of creation of the snarls, and the twist given to the yarn during the binding process. An increase in the value of linear density causes higher stiffness, which simultane-ously decreases the ability of creating snarl on the yarn. This also suggests a high unevenness of the linear density.
- There is a linear dependence ($R^2 = 0.962$ at a level of significance of 0.01) between the air index of the yarn and its linear density. The higher the linear density, the lower the ability of the air to carry the yarn through the nozzle of the air friendliness tester.

LOOP YARN

- The statistical analysis confirms that there is a statistical dependence between the angle of inclination of the binding yarn and the twist given to the yarn in the process of binding. The dependence is non-linear: the value of the angle in-creases (change of the angle from acute to straight) with the square value of the twist. This dependence demonstrates that the accumulation of braids on the length of the yarn causes a change in the direction of the angle. The determination coefficient R^2 = was 0.724 at a level of significance of 0.01.
- There is a statistical dependence be-tween the standard deviation from the average value of the angle of inclination of the binding yarn and the twist given to the yarn in the process of binding. The dependence is non-linear: the standard deviation from average of the value of the angle of inclination of the binding yarn increases with the square value of the twist. This dependence shows that the dispersion around the average value of the angle of the binding yarn inclination might be explained by an unevenness of the thickness of the core and effect components. For a low thickness of the components, the binding yarn covers them under a straight angle when the final twist is given. For a high thickness of the components, the angle of binding yarn inclination might be changed in the direction of the acute angle.
- There is a statistical connection be-tween the average value of the distance between loops and the twist given to the yarn in the final

process. The dependence is linear, with the determination coef-ficient R^2 = of 0.936 at a level of signifi-cance of 0.01.

- There is a linear dependence (determi-nation coefficient $R^2 = 0.983$ at a level of significance of 0.01) between the linear density and the length of the loop cre-ated at the end of the snarl effect. The higher the linear density of the yarn, the longer the loop at the end of the snarl. This might be explained by the increased stiffness of the yarn as a result of adding another effect component. In such a case, the ability of the yarn to create snarls with a small loop at the end decreases.
- There is a statistical dependence be-tween the elongation measured at the breaking moment and the twist given in the final process. The dependence is non-linear; the higher the square value of this twist, the higher the elongation value. The determination coefficient R2 = was 0.589 at a level of significance of 0.01.
- There is a linear dependence ($R^2 = 0.801$ at a level of significance of 0.01) between the air friendliness and the lin-ear density of the yarn. The higher the linear density value, the lower the ability of the air to carry the yarn through the nozzle of the air friendliness tester.
- The square value of the number of ef-fect components influences the standard deviation from the average value of the yarn carried through the nozzle of the air friendliness tester. This might be ex-plained by the increased number of pores in the yarn with many components. Not only is the relief of the yarn changed by adding another component, which may give a higher resistance to the air, but also the air caught in the pores makes the yarn heavier. The dependence is linear, with a determination coefficient $R^2 = 0.801$ at a level of significance of 0.01.

BUNCH YARN

- There is a linear dependence with determination coefficient $R^2 = 0.801$ at a level of significance of 0.01 between the width of the hysteresis loop obtained in the bending rigidity test and the twist given to the yarn in the final process of binding.
- There is a connection between the average value of bending rigidity and the twist given to the yarn in the binding process. The determination coefficient $R^2 = 0.903$ at a level of significance of 0.01.
- There is a linear dependence ($R^2 = 0.825$ at a level of significance of 0.01) between the standard deviation from the average value of the bending rigidity and the square value of twist given to the yarn in the final process: the higher the square value of this twist, the higher the standard deviation from the average bending rigidity value.

- There is a linear dependence (determi-nation coefficient R^2 = 0.869 at a level of significance of 0.01) between the angle of inclination of the binding yarn be-tween the bunches and the binding twist value: an increasing twist value causes a change of the angle of inclination from the acute to the straight direction.
- The frequency of the rocker's ac-tion influences the distance between the bunches. The higher the frequency of the rocker's actions, the smaller is the dis-tance between the bunches on the yarn. The determination coefficient R^2 = 0.866 at a level of significance of 0.01.
- There is an influence (R^2 = 0.866 at a level of significance of 0.01) of the fre-quency of the rocker's action on the stand-ard deviation from the average distance between bunches. The higher the fre-quency of the rocker's actions, the lower the standard deviation from the average distance between the bunches on the yarn.
- There is an influence (R^2 = 0.721 at a level of significance of 0.01) of the twist given to the yarn in the final process on the standard deviation from the average length of the bunches. The higher the value of the twist, the lower the value of the standard deviation from the average length of the bunches.
- The square value of the twist in the binding process influences the standard deviation from the average distance be-tween the clamps of the snarling tester at the moment of obtaining the snarl. The increase in the twist causes the average distance between the mentioned clamps to decrease. The determination coefficient R^2 = 0.735 at a level of significance of 0.01.
- The effect yarn overfeed influences the number of twists on the snarl. It suggests that periodical effects on the yarn such as bunches reduce the ability of the yarns to create long snarls. Bunch yarns with a higher overfeed of effect yarns, which means a thicker bunch, have a lower abil-ity to create snarls with many twists along them. The determination coefficient R^2 =0.500 at a level of significance of 0.01.
- The number of component yarns in-fluences the air index value. This might be explained in a similar way to the case of the spiral yarn (point 6). The determi-nation coefficient R^2 = 0.613 at a level of significance of 0.01.

TECHNICAL YARNS AND TEXTILES

Technical yarns are produced for the manufacture of technical textiles. They have to meet the specific functional requirements of the intended end-use. This may be achieved through special yarn production techniques or through the selection of special fibre blends or a combination of both. This chapter describes the yarn production technologies that are applicable to technical yarns and discusses the structures and properties of the yarns that may be produced using these technologies.

STAPLE FIBRE YARNS

Ring Spinning

Ring spinning is currently the most widely used yarn production method. Initially developed in America in the 1830s, its popularity has survived the emergence of much faster spinning technologies. In addition to the superior yarn quality, ring spinning is extremely versatile. It is capable of producing yarns with wide ranges of linear density and twist from a great variety of fibre materials. It is also used for doubling and twisting multifold and cabled yarns. Fibre materials must be properly prepared before they can be used on the ring spinning machine. The preparation processes are dependent on the fibre material. The typical process routes for cotton and wool. The ultimate objectives of the many preparation processes are to produce a feed material for the final spinning process that is clean, even, homogeneous and free from fibre entanglement. The fibres must also be in the preferred orientation.

On the ring spinning machine, the feed material is attenuated to the required linear density by a drafting system, typically a roller drafting system with three lines of rollers. The drafted fibre strand is then twisted by the ring spindle. The yarn leaving the front rollers is threaded through a yarn guide (the lappet), which is located directly above the spindle axis. The yarn then passes under the C-shaped traveller onto the bobbin. The bobbin is mounted on the spindle and rotates with the spindle. When the bobbin rotates, the tension of the yarn pulls the traveller around the ring.

During production, the bobbin rail moves up and down to spread the yarn along the length of the bobbin so that a proper package can be built. The movement of the ring rail is quite complicated, but the aim is to build a package that is stable, easy to unwind and contains the maximum amount of yarn. As the yarn is wound on the bobbin, the bobbin diameter increases steadily during production.

Ring-spun yarns have a regular twist structure and, because of the good fibre control during roller drafting, the fibres in the yarn are well straightened and aligned. Ring spun yarns therefore have excellent tensile properties, which are often important for technical applications.

The ring spinning system can be used for spinning cover yarns where a core yarn, spun or filament, is covered by staple fibres. This can provide yarns with a combination of technical properties. For example, a high strength yarn with good comfort characteristics may be spun from a high strength filament core with natural fibre covering. Other technical yarns, such as flame-retardant and antistatic yarns can also be made by incorporating flame-retardant and electricity conductive fibres.

The main limitation of the ring spinning system is the low productivity. The other limitations are the high drafting and spinning tensions involved. These high tensions can become a serious problem for spinning from fibres such as alginate fibres that have low strength.

FEED MATERIAL

Rotor Spinning

The productivity limitation of the ring spinning system was recognised long before the commercial introduction of rotor spinning in 1967. In ring spinning, the twist insertion rate is dependent on the rotational speed of the yarn package. This is so because of the continuity of the fibre flow during spinning. Numerous attempts have been made since before the end of the 19th century, particularly since the 1950s, to introduce a break into the fibre flow so that only the yarn end needs to be rotated to insert twist. Very high twisting speeds can thus be achieved. In addition, by separating twisting from package winding, there will be much more flexibility in the form and size of the yarn package built on the spinning machine. This increases the efficiency of both the spinning machine and of subsequent processes. Rotor spinning was the first such new technology to become commercially successful and it is the second most widely used yarn production method after ring spinning.

The principles of rotor spinning. The fibre material is fed into an opening unit by a feed roller in conjunction with a feed shoe. The feed material is usually a drawn sliver. An opening roller is located inside the opening unit and is covered with carding wire, usually saw-tooth type metallic wire. The surface speed of the opening roller is in the region of 25-30ms$-^{1}$, approximately 2000 times the feed roller surface speed.

This high speed-ratio enables the fibres to be opened up into a very thin and open fibre flow. The fibres are taken off the opening roller by an air stream with a speed about twice that of the opening roller. The fibres are carried by the air stream, through the fibre transportation tube, into the spinning rotor. The air speed in the transportation tube increases, owing to the narrowing cross-section of the tube, as the air reaches the exit point inside the rotor. This ensures that the fibres are kept aligned along the airflow direction and as straight as possible.

The exit angle of the fibres from the transportation tube is at a tangent to the rotor wall and the surface speed of the rotor is faster than the exit air speed, so the fibres emerging from the transportation tube are pulled into the rotor, keeping the fibres aligned in the direction of the fibre flow. The centrifugal force generated by the rotor forces the fibres into the rotor groove. Because of the high surface speed of the rotor, only a very thin layer of fibres, usually one or two fibres in the cross-section, is deposited in the rotor when the rotor passes the fibre exit point of the transportation tube. Many such layers of fibres are needed to make up the yarn. This doubling up of the fibres in the rotor is called back doubling.

Because the yarn is formed in an enclosed space inside the rotor, trash particles remaining in the fibres can accumulate in the rotor groove. This leads to a gradual deterioration of yarn quality and in severe cases yarn breakage. The cleanliness of fibres is more critical for rotor spinning than for ring spinning. In order to improve the cleanliness of the fibres, a trash extraction device is used at the opening roller.

As the twist in the yarn runs into the fibre band in the rotor groove, inner layers of the yarn tend to have higher levels of twist than outer layers. Fibres landing on the rotating fibre band close to the yarn tail, or directly on the rotating yarn arm when the yarn arm passes the exit of the transportation tube, tend to wrap around the yarn instead of being twisted into the yarn. These wrapping fibres are characteristic of rotor-spun yarns.

Rotor-spun yarns usually have lower strength than corresponding ring-spun yarns because of the poorer fibre disposition in the yarn. This is the result of using an opening roller to open up the fibres, of transporting the fibres by airflow, and of the low yarn tension during yarn formation. The wrapping fibres also lead to a rougher yarn surface. Rotor yarns have better short term evenness than ring-spun yarns because of the back-doubling action.

The main advantage of rotor spinning is the high production rate. However, because of the lower yarn strength, rotor spinning is limited to medium to course yarn linear densities. It is also limited to the spinning of short staple fibre yarns.

Friction Spinning

Friction spinning is an open end spinning technique. Instead of using a rotor, two friction rollers are used to collect the opened-up fibres and twist them into the yarn. The principle.

The fibres are fed in sliver form and opened by a carding roller. The opened fibres are blown off the carding roller by an air current and transported to the nip area of two perforated friction drums. The fibres are drawn onto the surfaces of the friction drums by air suction. The two friction drums rotate in the same direction and because of the friction between the fibre strand and the two drum surfaces, twist is inserted into the fibre strand. The yarn is withdrawn in the direction parallel to the friction drum axis and delivered to a package forming unit.

The friction drum diameter is much larger than the yarn diameter. The diameter ratio can be as high as 200. A high twisting speed can thus be achieved by using a relatively low speed for the friction drums. Owing to the slippage between the drum surface and the yarn end, the yarn takes up only 15–40% of the drum rotation. Nevertheless, a high production speed, up to 300mmin^{-1}, can be achieved. For a finer yarn the twist insertion rate is higher with the same drum speed so the delivery speed can be practically independent of yarn linear density.

Because the yarn is withdrawn from the side of the machine, fibres fed from the machine end away from the yarn delivery tend to form the yarn core while fibres fed from the machine end closer to the yarn delivery tend to form the sheath. This characteristic can be conveniently exploited to produce core-sheath yarn structures. Extra core components, filaments or drafted staple fibres, can be fed from the side of the machine while the fibres fed from the top of the machine, the normal input, form the sheath.

Unlike ring or rotor spinning machines that are produced by many manufacturers around the world, friction spinning machines are only currently made by Dr. Ernst Fehrer AG of Austria. The DREF 2 machine that has recently been upgraded to DREF 2000. The company also produces the DREF 3 machine that has an extra drafting unit on the side of the machine for feeding drafted staple fibres as a core component.

The fibre configuration in friction-spun yarns is very poor. When the fibres come to the friction drum surface, they have to decelerate sharply from a high velocity to almost stationary. This causes fibre bending and disorientation. Because of the very low tension in the yarn formation zone, fibre binding in the yarn is also poor. As a result, the yarn has a very low tensile strength and only coarse yarns, 100 tex and above, are usually produced.

The main application of friction spinning is for the production of industrial yarns and for spinning from recycled fibres. It can be used to produce yarns from aramid and glass fibres and with various core components including wires. The yarns can be used for tents, protective fabrics, backing material, belts, insulation and filter materials.

Wrap Spinning

Wrap spinning is a yarn formation process in which a twistless staple fibre strand is wrapped by a continuous binder. The process is carried out on a hollow spindle machine. The hollow spindle was invented by DSO 'Textil' in Bulgaria. The first wrap spinning machine was introduced in the 1979 ITMA.

The staple roving is drafted on a roller drafting system similar to those used on ring frames and is passed through a rotating hollow spindle that carries a binder bobbin. The rotation of the hollow spindle and the bobbin wraps the binder around the staple strand.

To prevent the drafted staple strand falling apart before it is wrapped by the binder, false twist is usually generated in the staple strand by the spindle. To introduce the false twist, the staple strand is not threaded through the hollow spindle directly. It is wrapped around either a twist regulator at the bottom of the spindle or part of the spindle top. The false twist also allows the staple strand to be compacted before the binder is wrapped around it. This improves the yarn strength. Two hollow spindles can be arranged one above the other to wrap the staple strand with two binders in opposite directions. This is used to produce special effect yarns with a more stable structure. Real twist may also be added to the yarn by passing the wrapped yarn onto a ring spindle, usually arranged directly underneath the hollow spindle.

Core yarns, mostly filaments, can be added to the feed. This can be used to provide extra yarn strength or other special yarn features. An example is to use this method to spin alginate yarns. Alginate fibres are very weak and cause excessive breakages during spinning without the extra support of core filaments.

A variety of binders can be used to complement the staple core or to introduce special yarn features. For example, a carbon-coated nylon filament yarn can be used to produce yarns for antistatic fabrics. Soluble binders can be used for making yarns for medical applications. Wrap spinning is highly productive and suitable for a wide range of yarn linear densities. Yarn delivery speeds of up to 300mmin$-^{1}$ are possible. Because the binder is normally very fine, each binder bobbin can last many hours, enabling the production of large yarn packages without piecing. Because the staple core is composed of parallel fibres with no twist, the yarn has a high bulk, good cover and very low hairiness. The main limitation of wrap spinning is that it is only suitable for the production of multicomponent yarns. The binder can be expensive, increasing the yarn cost.

Air-jet Spinning

Air-jet spinning technology was first introduced by Du Pont in 1963, but it has only been made commercially successful by Murata since 1980. Du Pont used only one jet, which produced a low strength yarn. The Murata system has two opposing air jets, which improves the yarn strength. The twin-jet Murata Jet Spinner. Staple fibres are drafted using a roller drafting system with three or four pairs of rollers. The fibres are then threaded through the twin-jet assembly. The second jet N_2 has a higher twisting torque than the first jet N1. Immediately after leaving the front drafting rollers, the fibres in the core of the yarn are twisted in the twist direction of N_2. The fibres on the edges of the drafted ribbon are twisted by the weaker N1 and wrap around the core fibres in the opposite direction.

Because the jet system is located between the front drafting rollers and the yarn delivery rollers, neither of which rotates around the axis of the yarn, the twist inserted by the jets is not real twist and after the yarn passes through the jet system, the core fibres become twistless. The yarn strength is imparted by the wrapping of the edge fibres. Because of the small jet dimensions, very high rotational jet speeds are possible. Although twist efficiency is only 6-12% because of the twist resistance of the yarn, delivery speeds of up to 300mmin^{1} are possible. In a further development by Murata the second jet is replaced with a pair of roller twisters. The principle of yarn formation is similar to the twin-jet system. The new machine, the roller-jet spinner, is capable of delivery speeds of up to 400mmin–^{1}. However, the yarn has a harsher handle.

Air-jet spinning is used mainly for spinning from short staple fibres, especially cotton and polyester blends. The vortex spinner, the latest addition to the Murata jet spinner range, was shown in ITMA 99 for spinning from 100% cotton.

The air-jet system can be used to produce core-sheath yarn structures by feeding the core and sheath fibres at different stages of the drafting system. Fibres fed in from the back of the drafting system tend to spread wider under the roller pressure and form the sheath of the yarn while fibres fed in nearer to the front tend to form the yarn core. Filament core can also be introduced at the front drafting roller. Two spinning positions can be combined to produce a two-strand yarn that is then twisted using the usual twisting machinery. Air-jet yarns have no real twist, therefore they tend to have higher bulk than ring and rotor yarns and better absorbency. They are more resistant to pilling and have little untwisting tendency. Because the yarn strength is imparted by the wrapping fibres, not the twisting of the complete fibre strand, air-jet yarns have lower tensile strength than ring and rotor yarns. The system is only suitable for medium to fine yarn linear densities as the effectiveness of wrapping decreases with the yarn thickness. The rigid yarn core of parallel fibres makes the yarn stiffer than the ring and rotor yarns.

Twistless Spinning

Numerous techniques have been developed to produce staple yarns without twisting so that the limitations imposed by twisting devices, notably the ring traveller system, can be avoided and production speed can be increased. Because of the unconventional yarn characteristics, these techniques have not gained widespread acceptance commercially, but they do offer an alternative and could be exploited to produce special products economically.

Most of these twistless methods use adhesives to hold the drafted staple fibre strand together. They can produce low linear density yarns at a high speed. The adhesives may later be removed after the fabric is made and the fibres are then bound by the interfibre forces imposed by fabric constraints. This type of yarn has high covering power due to the untwisted yarn structure. However, these processes mostly involve additional chemicals and require high power consumption. The yarns can only be used for fabrics that offer good interfibre forces.

As an example, the TNO twistless spinning method. In the system shown here, the roving is drafted under wet conditions, which gives better fibre control. An inactive starch is then applied to the drafted roving, which is also false twisted to give it temporary strength. The starch is activated by steaming the package that is then dried. The later version of TNO twistless system replaces the starch with PVA (polyvinyl alcohol) fibres (5-11%), which melt at above 80 °C to bind the staple fibres. This is also known as the Twilo system.

Another twistless spinning method is the Bobtex process, which can produce high strength yarns for industrial/leisure fabrics, such as tents, workwear and sacks. In this process, staple fibres (30-60%) are bonded to a filament core (10-50%) by a layer of molten polymer (20-50%). Production speeds of up to 1000mmin^{-1} can be achieved. The process can use all types of staple fibres including waste fibres.

Ply Yarn

Single yarns are used in the majority of fabrics for normal textile and clothing applications, but in order to obtain special yarn features, particularly high strength and modulus for technical and industrial applications, ply yarns are often needed. A folded yarn is produced by twisting two or more single yarns together in one operation, and a cabled yarn is formed by twisting together two or more folded yarns or a combination of folded and single yarns.

The twisting together of several single yarns increases the tenacity of the yarn by improving the binding-in of the fibres on the outer layers of the component single yarns.

This extra binding-in increases the contribution of these surface fibres to the yarn strength. Ply yarns are also more regular, smoother and more hard wearing. By using the appropriate single yarn and folding twists, a perfectly balanced ply yarn can be produced for applications that require high strength and low stretch, for example, for tyre cords.

A typical process route of a ply yarn involves the following production stages:

1. single yarn production
2. single yarn winding and clearing
3. assembly winding: to wind the required number of single yarns as one (doubling) on a package suitable for folding twisting
4. twisting
5. winding.

Twisting can be carried out in a two-stage process or with a two-for-one twister. In the two-stage process, the ring frame is used to insert a low folding twist in the first stage and an up-twister to insert the final folding twist in the second stage. The ring frame uses a low twist to enable higher delivery speeds. A suitable package is formed on the ring frame for the over-end withdrawal of yarn on the up-twister. The principle of the up-twister. The supply package rotates with the spindle while the yarn is withdrawn over the package end from the top. The free-rotating flyer is pulled around by the yarn and inserts twist in the yarn.

The two-for-one twister. The supply yarn package is stationery. After withdrawal from the package, the yarn is threaded through the centre of the spindle and rotates with the spindle. Each rotation of the spindle inserts one full turn in the yarn section inside the spindle centre and also one turn in the yarn section outside the yarn package (the main yarn balloon).

The yarn therefore gets two turns for each spindle rotation. If the supply package is rotated in the opposite direction of the spindle, then the twisting rate will increase by the package rotational speed. The Saurer Tritec Twister is based on this principle. In the Tritec Twister, the package rotates at the same speed as the spindle, but in the opposite direction, so each spindle rotation inserts three twists in the yarn. The package is magnetically driven.

The production of a ply yarn is much more expensive than the production of a single yarn of equivalent linear density. Not only does the ply yarn production require the extra assembly winding and twisting processes, but the production of the finer component single yarn is also much more expensive.

FILAMENT YARNS

Definitions

A filament yarn is made from one or more continuous strands called filaments where each component filament runs the whole length of the yarn. Those yarns composed of one filament are called monofilament yarns, and those containing more filaments are known as multifilament yarns. For apparel applications, a multifilament yarn may contain as few as two or three filaments or as many as 50 filaments. In carpeting, for example, a filament yarn could consist of hundreds of filaments. Most manufactured fibres have been produced in the form of a filament yarn. Silk is the only major natural filament yarn.

According to the shape of the filaments in the yarn, filament yarns are classified into two types, flat and bulk. The filaments in a flat yarn lie straight and neat, and are parallel to the yarn axis. Thus, flat filament yarns are usually closely packed and have a smooth surface. The bulked yarns, in which the filaments are either crimped or entangled with each other, have a greater volume than the flat yarns of the same linear density.

Texturing is the main method used to produce the bulked filament yarns. A textured yarn is made by introducing durable crimps, coils, and loops along the length of the filaments. As textured yarns have an increased volume, the air and vapour permeability of fabrics made from them is greater than that from flat yarns. However, for applications where low air permeability is required, such as the fabrics for air bags, flat yarns may be a better choice.

Manufacture of Filament Yarns

Most manufactured fibres are extruded using either melt spinning, dry spinning, or wet spinning, although reaction spinning, gel spinning and dispersion spinning are used in particular situations. After extrusion, the molecular chains in the filaments are basically unoriented and therefore provide no practical strength. The next step is to draw the extruded filaments in order to orient the molecular chains. This is conventionally carried out by using two pairs of rollers, the second of which forwards the filaments at approximately four times the speed of the first. The drawn filaments are then wound with or without twist onto a package. The tow of filaments at this stage becomes the flat filament yarn. The melt spinning process and the subsequent drawing process.

For many applications, flat filament yarns are textured in order to gain increased bulkiness, porosity, softness and elasticity in some situations. Thermoplastic filament yarns are used in most texturing processes. The interfibre bonds break and reform during the texturing process. A filament yarn is generally textured through three steps. The first step is to distort the filament in the yarn so that the interfibre bond is broken. Twisting or other means are used to distort the filaments within a yarn. The second step is to heat the yarn, which breaks bonds between polymers, allowing the filaments to stay crimped. The last step is to cool the yarn in the distorted state to enable new bonds to form between the polymers. When the yarn is untwisted or otherwise released from its distorted state, the filaments remain in a coiled or crimped condition. There are many methods for yarn texturing, including false-twist, air-texturing, knit-de-knit, stuffer box and gear crimp. Among these, the false-twist is the most popular method. The principles of the main methods of yarn texturing.

FILAMENT TECHNICAL YARNS

There have been many types of filament yarns developed for technical applications, such as reinforcing and protecting. The reinforcing technical yarns have either high modulus, high strength, or both. Yarns for protecting applications can be resistant to safety hazards such as heat and fire, chemical and mechanical damage. There are many types of technical filament yarns used in various applications, it is only possible, therefore, to list just a few yarns here that are popularly used in the development of some technical textile products.

Aramid Filament Yarns

Aramid fibre is a chemical fibre in which the fibre-forming substance is a long chain synthetic polyamide where at least 85% of the amide linkages are attached directly to two aromatic rings. Nomex and Kevlar are two well-known trade names of the aramid fibre, owned by Du Pont. Aramid fibres have high tenacity and high resistance to stretch, to most chemicals and to high temperature. The Kevlar aramid is well known for its relatively light weight and for its fatigue and damage resistance. Because of these properties, Kevlar

29 is widely used and accepted for making body armour. Kevlar 49, on the other hand, has high tenacity and is used as reinforcing material for many composite uses, including materials for making boat and aircraft parts. The Nomex aramid, on the other hand, is heat resistant and is used in making fire fighters' apparel and similar applications.

Aramid yarns are more flexible than their other high performance counterparts such as glass and Kevlar, and thus are easier to use in subsequent fabric making processes, be it weaving, knitting, or braiding. Care should be taken, though, as aramid yarns are much stronger and much more extensible than the conventional textile yarns, which could make the fabric formation process more difficult.

Glass Filament Yarns

Glass is an incombustible textile fibre and has high tenacity too. It has been used for fire-retardant applications and also is commonly used in insulation of buildings. Because of its properties and low cost, glass fibre is widely used in the manufacture of reinforcement for composites. There are different types of glass fibres, such as E-glass, C-glass, and S-glass. E-glass has very high resistance to attack by moisture and has high electrical and heat resistance. It is commonly used in glass-reinforced plastics in the form of woven fabrics. C-glass is known for its chemical resistance to both acids and alkalis. It is widely used for applications where such resistance is required, such as in chemical filtration. The S-glass is a high strength glass fibre and is used in composite manufacturing.

Glass filament yarns are brittle compared with the conventional textile yarns. It has been shown that the specific flexural rigidity of glass fibre is $0.89 mNmm^2 tex{\sim}^2$, about 4.5 times more rigid than wool. As a result, glass yarns are easy to break in textile processing. Therefore, it is important to apply suitable size to the glass yarn to minimise the interfibre friction and to hold the individual fibres together in the strand. Dextrinised starch gum, gelatine, polyvinyl alcohol, hydrogenerated vegetable oils and non-ionic detergents are commonly used sizes. When handling glass fibres, protective clothing and a mask should be worn to prevent skin irritation and inhalation of glass fibres.

Carbon Filament Yarns

Carbon fibres are commonly made from precursor fibres such as rayon and acrylic. When converting acrylic fibre to carbon, a three-stage heating process is used. The initial stage is oxidative stabilisation, which heats the acrylic fibre at 200-300 °C under oxidising conditions. This is followed by the carbonisation stage, when the oxidised fibre is heated in an inert atmosphere to temperatures around 1000 °C. Consequently, hydrogen and nitrogen atoms are expelled from the oxidised fibre, leaving the carbon atoms in the form of hexagonal rings that are arranged in oriented fibrils. The final stage of the process is graphitisation, when the carbonised filaments are heated to a temperature up to 3000 °C, again in an inert atmosphere. Graphitisation increases the orderly arrangement of the carbon atoms, which are organised into a crystalline structure of layers. These layers are well oriented in the direction of fibre axis, which is an important factor in producing high modulus fibres.

Like the glass yarns, most carbon fibres are brittle. Sizes are used to adhere the filaments together to improve the processability. In addition to protecting operatives against skin irritation and short fibre inhalation, protecting the processing machinery and auxiliary electric and electronic devices needs to be considered too, as carbon fibre is conductive.

HDPE Filament Yarns

HDPE refers to high density polyethylene. Although the basic theory for making super strong polyethylene fibres was available in the 1930s, commercial high performance polyethylene fibre was not manufactured until recently. Spectra, Dyneema, and Tekmilon are among the most well-known HDPE fibres. The gel spinning process is used to produce the HDPE fibre. Polyethylene with an extra high molecular weight is used as the starting material. In the gel spinning process, the molecules are dissolved in a solvent and spun through a spinneret. In solution, the molecules which form clusters in the solid state become disentangled and remain in this state after the solution is cooled to give filaments. The drawing process after spinning results in a very high level of macromolecular orientation in the filaments, leading to a fibre with very high tenacity and modulus. Dyneema, for example, is characterised by a parallel orientation of greater than 95% and a high level of crystallinity of up to 85%. This gives unique properties to the HDPE fibres. The most attractive properties of this type of fibre are: (1) very high tenacity, (2) very high specific modulus, (3) low elongation and (4) low fibre density, that is lighter than water.

HDPE fibres are made into different grades for different applications. Dyneema, for example, is made into SK60, SK65 and SK66. Dyneema SK60 is the multipurpose grade. It is used, for example, for ropes and cordage, for protective clothing and for reinforcement of impact-resistant composites. Dyneema SK65 has a higher tenacity and modulus than SK60. This fibre is used where high performance is needed and maximum weight savings are to be attained. Dyneema SK66 is specially designed for ballistic protection. This fibre provides the highest energy absorption at ultrasonic speeds.

YARN PROPERTY CONTROL ON A TWIST DRAFTING FRAME

Twist drafting has desirable properties for small-scale production of functional yarns. In twist drafting, a yarn is directly spun out from a large mass of fibers, and the withdrawal tension in the yarn is used to maintain steady-state conditions of yarn forming and yarn evenness. This work presents a mechanical analysis of twist drafting and a related experiment. The analytical results reveal the relation between the yarn properties of linear density and twist rate and the operational parameters for spinning of withdrawal tension in the yarn being spun out and the porosity ratio in the mass. The relation permits control of the linear density and twist rate of the yarn independently to a certain extent and systematically, which is impossible with conventional twist drafting.

Fig. twist drafting machine

Twist drafting produces a yarn directly from a large mass of material fibers, without dealing with some semi-manufactures such as sliver and roving, by using the twist of the yarn to supply the nip force to select fibers from the mass. Twist drafting has some desirable properties for small-scale production of various functional yarns: ability to process extremely small lots, high adaptability to low spinnability materials, simplicity of the pretreatment process, and low frame-making costs due to the simple frame mechanism. The literature lists several studies on functional yarns spun by twist drafting frames (TDF). Matsumoto et al. developed an improved throstle-spinning machine for a silk/raw silk core-spun yarn. In earlier work, we successfully examined a functional yarn made from a carbonized cotton/polyester blend with tension control twist draft spinning (TCTDS).

In that work, we presented the features and properties of TCTDS, a kind of TDF, along with a mechanical analysis of twist drafting based on Hannah's theoretical work. The results of our analysis showed that the product of the diameter and twist rate of a yarn being spun out becomes a constant, which is determined by its fiber properties-length, diameter, and coefficient of interfiber friction. This analytical result happened to theoretically reveal a defect of conventional twist drafting: once a material fiber is selected, the diameter and twist rate of the yarn depend on each other. Moreover, this analytical result is not a sufficient tool for more practical design of TDF yarns, where specifications of linear density and twist rate of the yarn are given. No one has succeeded in revealing theoretically which variables affect the linear density and twist rate.

This paper presents our subsequent mechanical analysis of twist drafting, especially the linear density and twist rate of yarn, along with a related experiment using flyer-type twist draft spinning (FTDS) (Figure 1), a kind of TDF. Moreover, we suggest a technique based on the analysis for independent control of both the linear density and twist rate of TDF yarn.

FTDS

Spinning with FTDS (Figure 1), a kind of TDF, is summarized as follows: a yarn is spun out directly from a large treated mass of fibers in a pot, is twisted by the rotation of a flyer, and is wound up by a bobbin. Yarn twisting and drafting are performed simultaneously. Continuous twisting at a sufficiently high speed gives more twist to the yarn, leads the number of fibers to be selected and drawn out by the yarn from the larger mass, and therefore leads to larger withdrawal tension and thickness of the yarn. The inverse phenomenon occurs when twisting stops.

Taking this fact into account, the flyer rotation on-off control system operates the twisting speed so as to make the withdrawal tension constant, so it is able to maintain yarn evenness. The reason nearly uniform yarns are attained by keeping the withdrawal tension constant is that the number of fibers drawn out from the mass is approximately proportional to the tension if the density of the mass is invariant.

The rotational speed of the bobbin is determined depending on the twisting speed of the flyer and the amount of yarn take-up, keeping the yarn take-up speed constant.

The pot has the function of adjusting the density of the mass, i.e., 1-[epsilon] ^ sub t0 ^, where [epsilon] ^ sub t0 ^ is the porosity ratio in the mass. Detailed diagram of the pot, which is composed of a holder and a pressure adjusting board that moves up and down in parallel with the bottom face and along the inner wall of the holder and can be fixed to the holder. The mass is sealed by being pinched between the holder and the pressure adjusting board. For that reason, 1-[epsilon] ^ sub t0 ^ can be adjusted efficiently and precisely by adjusting the fixing position of the pressure adjusting board. We prepared the mass as follows:

First, we made a fleece by putting the material fibers on a carding machine. Second, we split the fleece in the direction of the fibers according to the width of the holder. Third, we packed the split fleece uniformly in the holder, and pinched it with the pressure adjusting board.

INTEREST OF A COMPOUND YARN TO IMPROVE FABRIC PERFORMANCE

Textiles in many forms are used in interiors of public transportation (buses, aircrafts, cars, train and so on). These textile structures have to respect transportation legislation which is very strict regarding personal safety. However, the flame-retardant treatments suitable for these mainly wool-containing fabrics are generally not permanent after dry cleaning and laundry treatments 1.

At the same time, the textiles used for these applications require other specific properties such as dyeability, good handle etc.This is why wool is particularly used as seat fabric in the interiors of buses and cars for its comfort and coloration properties, coupled with its low level of inherent flame resistance. However, the market increasingly requires highly flame- and heat-resistant textiles (such as seat coverings in aircraft) to have a combination of other acceptable properties including good handle. The knitted articles obtained with compound yarns are studied with regard to different properties such as resistance to fire, cutting, dyeing and abrasion. The effect of UV on their behaviour has been analysed.

EXPERIMENTAL

Materials

Poly-p-phenylenediamine-terephtalamide fibres (PPTA fibres) are classical Kevlar®29. The compound yarn will further be denoted as DREF yarn. The fibres used have the characteristics described in Table 1. The diameter of the wool fibres was determined using a Fibre Fineness Meter (Air Flow) [2], and the linear density of PPTA fibres by using a Vibroskop [3]. We measured the fibre length on a Wira Fibre Diagram Machine

Spinning Process

In this study, we produced DREF yarns with a count of 71 tex and with variable amounts of PPTA fibres added (20, 30 and 50 wt.-%). The spinning principle we used is a non-conventional spinning process called DREF [5]. The DREF 2000 apparatus is composed of a rotating carding drum which opens the slivers into single fibres, and a specially designed inlet system being used for sliver retention.

The fibres are then stripped from the carding drum by centrifugal force and carried into the nip of the two perforated spinning drums. The fibres are subsequently twisted by mechanical friction on the surface of the drums, which rotate in the same direction (S or Z). The process is assisted by air suction through the drum perforations.

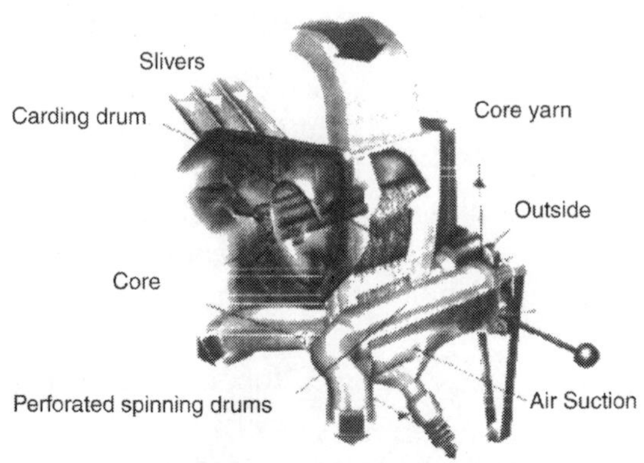

Fig. DREF Princile

The fibres, which form the slivers used in this process, had a maximum length of 60 mm. The experimental conditions were as follows: production speed of 60 m/min, spinning drum speed of 2500 rev/min and pressure of -45 mbar. The yarn obtained was a core yarn. The core was a Kevlar yarn, coated by wool fibres to protect the Kevlar against UV.

Knitting process

The DREF yarns were knitted on an automatic rectilinear machine gauge 7 – Shima Seiki SES 122 FF. The texture used was a double woven rib (the samples had a specific weight of 1.0 ± 0.05 kg/m^2).

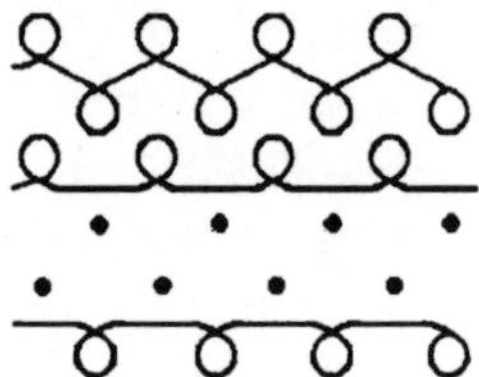

Cutting resistance

The cutting resistance of the knitted articles was tested on a COUPTEST [6] according to the AFNOR standard EN 388-1994. A cutting index "I" was obtained. Each sample was tested 10 times in two different sequences of 5 tests.

Cone Calorimetry by Oxygen Consumption

The samples were exposed to a Stanton Redcroft Cone calorimeter at a heat flux equalling 75 kW/m^2. This flux corresponds to flashover conditions chosen because of the high fire performance of our fabrics. With a lower heat flux, the samples would not have burned. The samples were placed between two cut steel sheets. The surface exposed to the external heat flux measured

9×9 cm². Our method did not correspond to any standard. The conventional data obtained were Rate of Heat Release (RHR) and Volume of Smoke Production (VSP) [9], using software developed in our laboratory. The experiments were repeated 3 times. When measured at 75 kW/m² flux, RHR and VSP values were reproducible to within ±10%. The cone data reported in this work wasis the average of three replicated experiments.

Cutting Resistance

The DREF yarns had variable amounts of PPTA (20, 30 and 50 wt.- %), they were knitted on a double woven rib given in Figure 2. The fabrics then made had almost the same specific weight and thereforecould be compared.

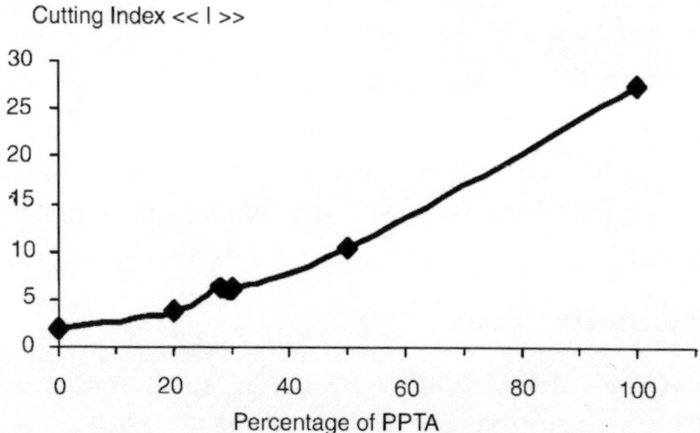

Fig. Cutting Performance of DREF Yarns

Figure 3 shows the cutting behaviour of the knitted fabrics. We observe that the cutting index is about 10 for the samples wool/PPTA 50/50, which represents a very good potential. The standard establishes a classification in 5 levels (1 to 5) and the passage from class 4 to 5 is equivalent to an increase cutting index of 20 [10]. In this case, only Kevlar® could pass to the 5th class. The sample in 50/50 wool/PPTA exhibited a cutting resistance of close to the 4th level, followed respectively by wool/PPTA 70/30 (3rd level), wool/PPTA 80/20 (2nd level) and wool (1st level).

Flame Resistance of DREF Yarns

The fire performance of the wool/PPTA blends increased as the amount of PPTA increased (Figure 4). It can be observed that the RHR peak of wool/ PPTA 50/50 blend is slightly higher than the one obtained with PPTA alone, but the time to ignition is lower than with PPTA alone and corresponds to the one obtained with wool. The RHR curves of the blends exhibit a double RHR peak, corresponding to the peaks of wool and PPTA respectively [12]. The last RHR peaks always occur at the same time (about 25 s) whatever the composition of the blend. This result suggests that interactions take place between wool and PPTA during fire degradation.

Dyeing Tests

The advantage of DREF yarns (core yarn), is the covering of PPTA by wool. The wool protects PPTA against UV (PPTA is easily degraded by UV), and it is possible to dye the wool with colours and design according to the requirements of the manufacturer.

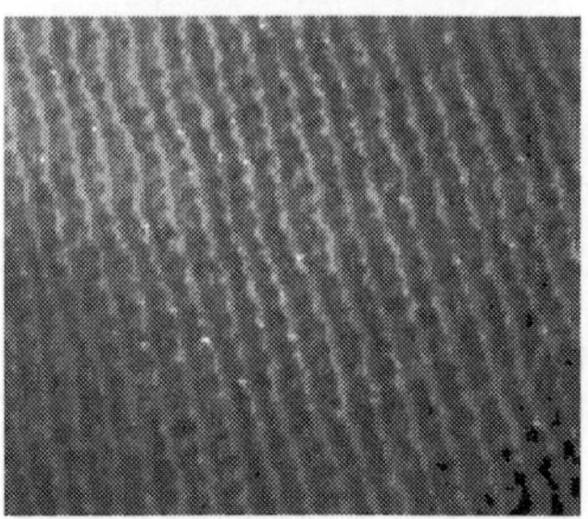
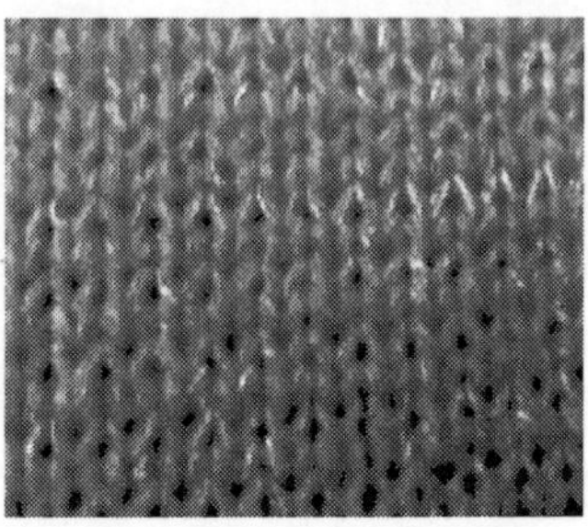

Fig. Dyeing Samles Woo/PPTA 88/20 and wool/PPTA50/50

The knitted articles were dyed with an acid dye. We may note in Figure 5 that we obtain a good dyeing of fabrics, irrespective of the percentage of PPTA, although some yellow spots are visible due to bad wrapping of PPTA fibres.

Spectro-photocolorimeter Tests

Spectro-photocolorimeter tests were carried out on knitted articles before and after ageing in a Xenotest apparatus, using a xenon-arch lamp as a light radiation to simulate natural sunlight. The samples were exposed for 100 hours. The results show that the loss of colour shade is reduced by 3 times for knitted structures with DREF yarns in comparison to a PPTA fabric.

YARN TENSION IN THE PROCESS OF ROTOR SPINNING

Yarn tension is a phenomenon inseparably connected with spinning, and is also of fundamental importance during rotor yarn formation. The problem of the occurrence of dynamic yarn tensions is of essential importance, as the possible decrease in yarn quality parameters, such as irregularity of linear density, elongation, tenacity, and the number of faults, may result. Consideration of real spinning conditions allows us to state that the temporal tension values are more important than the mean values.

Investigation into the dynamic forces which act on rotor-spun yarn carried out by Grosberg and Mansour confirmed the occurrence of high-frequency (short-wave) yarn tension oscillations of significant amplitudes, which even exceed the mean value of about 30 %. Maximum yarn tensions directly influence the number of yarn breakages, which in turn decrease the spinning machine's efficiency and raise yarn manufacturing costs. The distinct influence of yarn tension is visible when analysing the stretching forces of yarn. An increase in yarn tension during spinning causes a decrease in the elongation of yarn over stretching to break. This influence is very substantial, although the yarn tension values constitute only 10% -20% of the yarn's breaking force.

Fig. Rotor Spinning Machine

Tests carried out hitherto showed that spinning tension is determined by the rotor's rotational velocity and its diameter. Further experiments and considerations (conducted at the Department of Technology and Structure of Yarns) allowed us to draw the conclusion that the coefficient of variation of yarn tension is well correlated with the coefficient of variation of linear density of the sliver feeding the spinning machine, the coefficient of variation of yarn linear density, and the coefficient of variation of yarn tenacity. From investigations carried out at the Department of Technology and Structure of Yarns at the Technical University of £ódŸ, it also appears that the mass distribution of the fibre stream feeding the rotor spinning machine has an essential influence on the irregularity of the yarn's linear density, and on the distribution of temporal tensions during spinning. Every mass change in the feeding fibre stream causes an increase or decrease in the irregularity of the yarn's linear density, and these phenomena are reflected by yarn tension changes. Notwithstanding the numerous investigations and theoretical elaborations carried out, no model which would include all the parameters of the spinning process and would link these parameters with yarn tension has so far been developed. The models so far elaborated have only considered the influence of the rotor's diameter and velocity and of the yarn's linear density, without considering the quality parameters of the product which feed the spinning machine. The influence of the sliver's linear density was investigated only by an experiment under real conditions. All these considerations motivated us to undertake an attempt at developing a model which would represent the phenomena occurring during the process of rotor yarn formation as closely as possible. This model should also take into account (in a generalised way) the influence of the following parameters on yarn tension during the process of rotor spinning:

ÀÛƒÜ the rotor's diameter and velocity,

ÀÛƒÜ the linear density of yarn, and

ÀÛƒÜ the quality parameters of sliver feeding the process.

SUBJECT OF INVESTIGATION

Within the scope of the investigations carried out, tests were performed which were aimed at estimating the influence of the yarn's linear density on the tension value existing during the spinning process. Cotton yarns with linear densities of 25, 35, 45, and 55 tex formed with the use of BD 200S rotor spinning machine were used for tests. The analysis of the quality parameters of the yarns formed was carried out by means of a computer measuring system developed within the scope of the research project, and realised at the Department of Technology and Structure of Yarns in co-operation with the Department of Applied Information Technology at the Faculty of Electricity and Electronics.

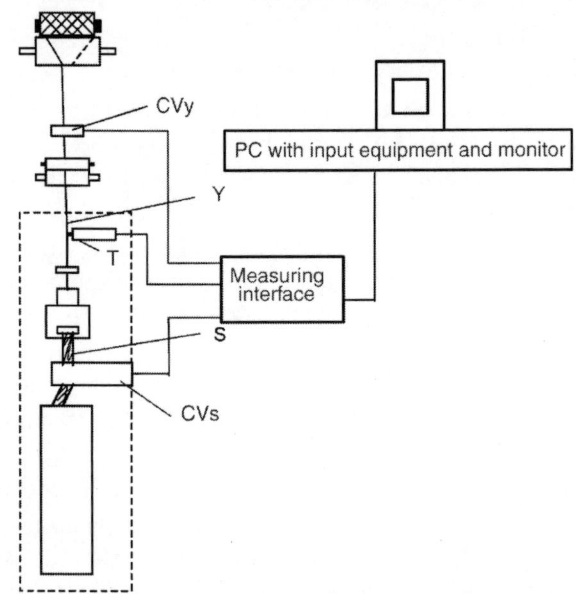

Fig. Computer Measuring system; T-yarn tension measuring head, CVs-siliver linear density measuring head, Y-yarn, S-sliver

The system allows continual recording of the following fundamental parameters of the spinning process:

- Linear density of yarn,
- Linear density of sliver,

- yarn tension

The basic elements of this system are a computer and three measuring heads connected with the computer by means of a measuring interface. The T-measuring head (based on a tensometric gauge) measures yarn tension, whereas the two USTER electro-capacity measuring heads were designated for determining the linear density irregularity.

WOOLEN OR WORSTED-THE DIFFERENCE

Woolen and Worsted are two major classifications for wool yarns and fabrics

Woolens	Worsteds
Processing	
Spun from shorter wool fibers-one to theree inches in length spum from fibers of a medium or course diameter fibers are washed, scoured and carded.	Spun from longer wool fibers-longer than three inches in length spun fibers of fine fimeter fibers are washed, scoured, carded, combed and drawn.
Yarn	
Bulky, uneven low to medium slack twist tensile strength lower than worsted.	Fine, smooth, even tighter twist higher tensile strength
Fabric Appearance	
soft fuzzy thick, heavier weight	Crisp smooth, clear-faced lighter weight
Characteristics more insulatory due to trapped air not as durable as worsteds nap reduces shine does not hold a crease well.	Less insultory more durable than woolens may become shiny with use where abraded during wear holds creases and shape.
Uses	
Sweater, carpets tweeds	Suits, dresses gabardines, crepes

The Woolen System

Woolen fabrics are characterized as being fuzzy, thick and bulky. They are made from fibers that are one to three inches in length that have been carded only (worsted yarns are carded and combed). After the carding process the woolen "sliver" is twisted by machine into rope-like strands called "roving" and wound onto reels for spinning. Woolen yarns are fluffy, relatively loosely twisted, and are used in weaving fabrics such as tweeds, blanket cloth and meltons. Woolen fabrics and yarns are traditionally made into bulky garments such as coats, heavy jackets and sweaters.

The Worsted System-combing and Drawing

Worsted yarns are spun from longer (three inches and longer) fibers that have been carded, combed and drawn. Combing machines further straighten the wool sliver making the individual fibers lie parallel. The combing process also eliminates "noils" or shorter fibers which grow mostly on the belly of the sheep. (Noils are used in the production of less expensive woolen fabrics and for the manufacturing of felt, a non-woven fabric).

The drawing process takes the worsted sliver, doubles it over onto itself and draws it out again to a thinner, more uniform diameter to insure that all wool fibers are parallel. Worsted yarns are twisted tighter and thinner in the spinning process and are manufactured into lightweight fabrics such as gabardine and crepe.

WOOL IN YOUR WARDROBE

When planning a professional business wardrobe, wool is preferred. Chosen for its comfort, appearance and longevity, wool garments are suitable form many situations and individual styles. Because wool has this versatility, it is both prestigious and practical for a wardrobe. When concerned with the price-value relationship of clothing, look closely at clothing durability and appearance over time. Authorities (consultants) on clothing selection and image consistently recommend selecting fewer, but higher quality clothes. By selecting garments qualitatively, rather than quantitatively, money is saved in the long run and a better investment has been made. The right business, professional or social image is important. It can add to your confidence and credibility. A wardrobe that contains wool is attractive, resists wrinkling, tailors beautifully, retains its shape, wears well, travels well and projects an effective professional image. A wool garment will give the appearance needed to get through a long, busy day.

Today's lifestyles demand getting the best value for your time and money. Make wool a financial investment, and watch it pay off.

CONSUMER ATTITUDES TOWARD WOOL

WARM, SOFT, COMFORTABLE, NATURAL, PURE: The words used by consumers to describe wool in a research study conducted by Bruskin Goldring Associates. The study entitled, "Consumer Attitudes Towards Wool and Wool Products," asked consumers what they look for in wool garments. The top four answers were comfort, fit, durability, and ease of care. This research bears out the growing trend to comfort-driven and quality of clothing. After years of living in T-shirts and sweats, the American consumer has become accustomed to clothing that is easy on the body. When Americans dress up, they are still likely to be slightly dressed-down.The boom in knitted apparel is just one indication of the strength of this trend. Layered-knit dressing, comfort-stretch wovens, and the return to sweaters all point to the importance of the simple but chic style that wool fits so well. The American wool industry is responding to these demands with innovative blends and finishes. Fabric weights are generally lighter than they were even a few years ago. Feather-weight worsteds and twelve-ounce coatings are just two of the wool fabrics that have inspired designers to create easy-to-wear yet elegant garments.

BLENDS Yarns and fabrics made with blends of American wool and other fibers, either natural or synthetic, have long been the choice of knowledgeable designers. Blending two great natural fibers like wool and cotton produces fabrics which benefit from the inherent qualities of each fiber. The end result is better comfort, better aesthetics and better performance.

The American Wool Council has been involved in the development and promotion of wool/cotton blended yarns and fabrics for several years. ALAMERA is a registered trademark of the American Wool Council denoting blends of American wool and cotton. ALAMERA blends are distinguished by a soft hand and good loft. The fabrics are season-spanners which can be worn nine to twelve months out of the year, depending on the climate. Other current popular blends are wool/polyester, wool/rayon, wool/nylon and wool/silk.

FANCY YARNS

Fancy yarns in contrast to the ordinary plain yarns, feature regular or random variations in yarn structure that are immediately evident to the eye. They are used for ornamentation to increase the aesthetic appeal of the fabric and resultant garment. There are many hundreds of fancy yarns; all of them different to one another and offering the designer a wide variation and may be composed of various fibre types or blends.

The most common of these fancy yarns are the following:

Boucle

Curled Looped Yarns

Commonly, a threefold yarn which has distinctive loops obtained by twisting a heavy yarn around a foundation yarn which is slack in tension. Usually, two or three colours are used to highlight the looping or curling characteristic of the outer yarn. Boucle yarn was originally intended to imitate Astrakhan fur, whereas curled and loop yarns are of the same character but smaller in size.

Cable Yarns

These are normally of a plain nature with a very bulky surface. Usually, consist of six individual yarns twisted around one another. A further variation which is more subtle is to have three, twofold yarns again twisted around one another. The resultant yarns are much stronger and are used for cord effect in woven fabrics.

Chenille Yarns

These are a long pile yarn, normally made from 100% cotton and used for dressing gowns, bedspreads, cushion covers, etc. Variation of chenille yarn is used for rugs and carpets and is normally composed of wool rather than cotton.

Core Yarns

These yarns consist of two different types of fibres; one is spun into a yarn which forms the centre core around which then spun another fibre. Some of the most common of these are nylon or polyester core around which cotton is spun. Typical example of use for this type of thread is in the sewing industry in yarns for industrial use.

The basic idea is that the cotton protects the heat sensitive nylon or polyester fibre as it is being used in the sewing machine, for very high temperatures are reached when sewing machines are used at high speeds, particularly where the yarn passes through the eye of the needle. Cotton dissipates the heat so caused. If this were not used, the polyester or nylon would soften or melt. They are also used for tyre cord for special tyre cord fabric.

Crepe Yarns

These can either be single or twofold yarns. They are highly twisted to the extent that the high twist forms a kink in the yarn and tends to make a buckle if it is not held under tension, therefore, the fabric is woven with tension in both the warp and weft direction and on completion of the fabric when it is removed from the loom and relaxes, a pebbly surface effect is obtained.

Diamond or Chain Yarns

These are threefold yarns composed of thick soft yarn which is folded with another yarn of medium twist. This resultant twofold yarn is then twisted further with a fine, hard twisted yarn, and the twist being in the opposition direction to that which the first yarn was spun. Therefore, it is the finer yarn, usually of a different colour, that produce a diamond pattern on the base two-fold yarn.

Marl Yarns

These were originally woollen yarns made by blending two different colour rovings during spinning. These give the effect of a twofold yarn when in fact it is only singles yarn. This same system is now applied to cotton and synthetic fibres as well as the original woollen yarns for which it was designed.

Melange Yarns

Another worsted yarn, which was originated by printing a combed sliver in single or multi- coloured stripes across its length. Alternating deepening and fading of its colour characterize yarns. A popular yarn for both suitings and for rugs.

Botany Yarns

A worsted yarn spun from 60s quality. The term refers to the quality of the top, i.e. 60s or finer from which it was spun. The yarn is used for the manufacture of top quality suitings.

Fingerings

These are worsted yarns spun from fine crossbred wool usually two, three or four ply. Each ply is a different colour. They are used in hose and knitted goods.

Slub Yarns

The random spaced thick soft twisted portion in the yarns characterizes these. These slubs must be randomly spaced otherwise the effect is not achieved on the finished fabric. Slub yarns are used to weave fabrics to imitate linen fabrics or raw silk fabrics, e.g. shantung, cotton and rayon fibres are the most common used.

Texturised Yarns

Texturised yarns have greater bulk or to put it in another way, occupy a greater space or volume than the equivalent conventional yarn. They are normally spun from man-made fibres, particularly the synthetic type due to their thermoplastic property, which permits them to be texturised during the actual spinning process of the fibre during manufacture. The word texturised refers to a group of yarns produced for special effects. Please read pages 237-242 in Textiles for Modern Living for a fuller explanation of this very important topic.

FANCY YARNS: EFFORTS TO METHODISE, PROBLEMS, AND NEW SUGGESTIONS

Fancy yarns are special products of spinning, twisting, wrapping, texturing, printing, knitting, etc. Fancy yarns are and will always be up-to-date, as there is no alternative to them. The demand for yarns with structural and/or optical effects is due to the special aesthetic and high decorative appeal to the woven, knitted materials, and other textiles as well. Textile materials that are produced using yarns with effects find applications in normal and high fashion clothing. Such yarns are also used for decorative textiles like curtains, carpets, "effect" wallpaper. Decor materials and textile fabrics in the corporate sector, as for example in the trim of a car or textile furnishing of a hotel lobby are becoming more and more important.

Fancy yarns have deliberately introduced irregular characteristics, in either diameter and bulk and/or in color, etc. as well as virtually new structures composed of fibers, yarns or other products that differentiate them from conventionally spun or multifilament yarns. New looks, structures and raw materials of fancy yarns are constantly in demand.

Components of fancy yarns can be of natural or man made fibers: staple length or continuous filaments but nowadays it appears the especially unusual structures of fancy yarns that are very desirable in designing of new textiles. There are so many different types of fancy yarns. Most of them are produced using spinning or twisting machinery that are modified or specially developed for this purpose. Some types of such yarns are produced from "fancy fiber" or "fancy slivers" used as minor components of yarns made by spinners with normal equipment, still others are made exclusively by filament yarns using adaptations of various processes as well as tape-knit fancy yarns are produced using a range of take-up rations and component yarns on a knitting machine specially designed for fancy yarn manufacture.

In spite of the wide use of fancy yarns the grouping problems of such assortment are almost unsolved. Particularly it is a lack of date necessary to analyze the huge variety of assortment of fancy yarns. Up to now there were only some efforts [3 – 6] to propose the ways for the grouping of fancy yarns and such investigations are in want of validity of the suggestions. The available reference sources and data are scanty, and by no means homogeneous.

The purpose of the study reported in this paper is to try to fill this gap; it presents the analysis of existing ways of fancy yarns' grouping, discusses the new suggested systematization of all types of fancy yarns.

PECULIARITIES OF ASSORTMENT AND PROBLEMS OF ITS GROUPING

The first fancy yarns types became an established textile novelty considerably earlier but traditional fancy yarns experienced booming development thanks to the mechanical and technological progress. It made it possible to overcome production difficulties that are considerably higher than those of classic flat yarns did.

In general case fancy yarns have a multithread structure composed of three components: core, effect, and binder. The effects like loops, waves, knots, snarls, etc. are formed by the effect yarn which unites with the core. These two components form the effect intermediate product that is fastened by the binder component that fixes effects.

The search for novelties turned into reality new structures of fancy yarns. Their effect intermediate product consists of three yarns, four or more ones. In every year and even in every season there is run for something new, for the exclusive and particular yarn with high-fashion multi-colour compound effects. A vital part of the effect systems today is the microprocessor control, which allows problem-free production of any imaginable fancy yarn pattern, and taking care of all control, operating and monitoring functions.

Loop yarn is one of the kinds of fancy yarns with periodic loop effects. The geometry of the effect could be very different: the loops are closed or opened and of regular or irregular shape as well. Frotte yarn is a type of loop yarn but it is an important difference that marks out this type of yarn from the group of loop yarns. According to, there are two core yarns in the loop yarn, while in the frotte yarn there is only one core yarn, so this structure is rather different.

The spiral fancy yarn is a plied yarn with smooth spiralling of one component around the others. Delivering one or more of its components at a greater speed can produce fancy yarn with spirals, so the shorter length of the component forms the core, while the greater length of the components creates the spirals.

Knop yarn contains prominent bunches of its component arranged at regular or irregular intervals. Using rollers capable of being operated independently usually makes this fancy yarn: core component – intermittent delivery and effect component – continuous delivery. So the effect component joins the core and is gathered into bunch or knop by the insertion of twist. The type of yarn is characterised by a thickening of its diameter, called a slub. Such fancy yarn can be manufactured on the woollen, semi-worsted, or worsted spinning system. Spun slub fancy yarn may be produced by an intermittent acceleration of one pair of rollers during spinning.

The special fancy effect of rotor spun slub yarn is characterised by the long slub and this effect is quite different from ring spun slub.

Plucked slub fancy yarns are composed of two core yarns and periodic short lengths of materials that have been plucked from the roving.

The puff fancy yarn has an intermediate appearance between that of a slub yarn and knop yarn. The effect is extremely soft, just like a very supple slub. The formation of such effect is achieved because of special grooves in the feed rollers, so the fibres nipped by the draft rollers can easily include the other fibres. This allows forming extremely soft thickening.

Snarl yarn is produced using twist-lively component. In such component the snarl formation occurs at a temporary decrease in tension or is due to a yarn's slack. The shape of the effect depends on the twist level of the yarn and other characteristics as well.

There are many types of fancy yarns with optical effects. Such effects are colour and lustre/matness. Fibres of different colour are mixed at blow room or carding, or drawing stage. There are units that help to distribute multicoloured knops of felted fibre and to produce short colour effects – distinct and prominence or slight and flat points on the surface of spun fancy yarn. There are special equipment that make colour effects in the sliver. Components of different colour can form a multicolour effect, which is further drafted during the following spinning operations. Formation of colour effects can be achieved by means of printing or special dyeing of sliver, rowing, and yarn. Mouline yarn is made from two or more components coloured in different colours and plied together.

Colour effects frequently combine with the structural ones. Such fancy yarns are very decorative and attractive.

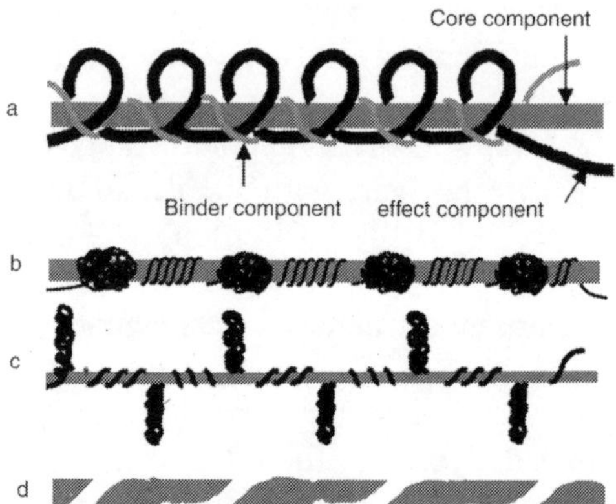

Fig. Some types of traditional fancy yarns with various effects: a- common view of loop fancy yarn, b, c, d-knot, snarl, wave effects

The technology of differential dying of the yarns or semi-manufactured product of spinning is effective and simple in design. Colouring and decorating the product using differential dyeing or cross dyeing techniques essentially promote the fibre properties as design effects in themselves without any other additional components.

Textile designers when creating new styles and fashionable textiles prefer differing from traditional fancy yarn structures the tape yarns or ladder-knit yarns. Such fancy yarns create extraordinary effects in the fabric made, thus augmenting their decorative value. Ladder-knit fancy yarn can be produced on a small-diameter circular knitting machine with three guides feeding each knitting head. Some knitted fancy yarns have effect component – short or long lengths of yarn, bunches of roving, etc. and colour effects as well.

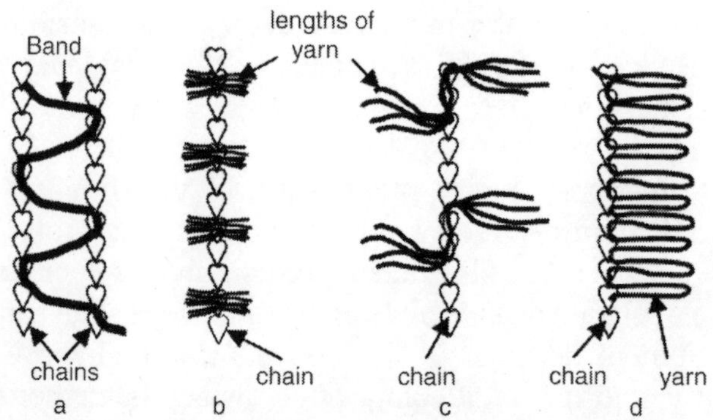

Fig. Some examples of not trditional fancy yarn structures:

a- ladder-knit fancy yarn,

b- knit fancy yarn with effect-short horizonatal lengths of yarn,
c- knit fancy yarn with effect-long lengths of yarn ("flag" effect),
c- knit fancy yarn with yarn effect

The cords are very decorative knitted fancy yarns. They are widely used in passementerie, knitting, trimmings, etc. Latch needles are employed and yarns of any kind can be processed: thus, it is possible to obtain chainette cords with many original effects of colour and of yarn couplings as well. Research papers discussing fancy yarns have predominantly described the assortment, the components and the techniques used. However, there are limited studies about the possibilities to methodise various fancy yarn types.

In the assortment of fancy yarns is grouped into two groups: with overfeed and controlled effects. Each of these two groups has two subgroups: yarn effects and spinning effects. For example, the overfeed yarn effects are loop, onde, snarls, wild knop, wild caterpillar, etc. The controlled yarn effects are knop, caterpillar, reverse caterpillar, and combined effects. Overfeed spinning effects are onde, spun loop, chenille, boucle, etc. Unfortunately, such a method of grouping is restricted and not precise because not all types of the assortment are included. Besides that, there are no motives of ascription and substantiation. Regular and controlled effects are separated. As further the assortment is grouped into fancy yarns with yarn effects and spinning effects as well. Regular and controlled effects are separated. As further the assortment is grouped into fancy yarns with yarn effects and spinning effects as well. In the classification of fancy yarn profiles capable of production on the hollow spindle fancy-wrap system is discussed. Different kinds of boucle fancy yarns are presented.

There are considerable imperfections in the above mentioned publications:

- Not all traditional types of fancy yarns are included,
- New structures of fancy yarns aren't included,
- Majority of fancy yarns' types aren't characterised,
- It's exists the confusion and vague meaning in the appellations of effects and groups of fancy yarns.

Possibilities to Methodise Fancy Yarns and New Suggestions

Analysis of the assortment enabled to determine the criteria that are significant for the systematisation of fancy yarn assortment:

- Method of manufacture,
- Use of fancy yarns and raw material of components,
- Number and type of components,
- Type of the effect,
- Material for effect making.

There are different technologies for fancy yarn manufacture. The choice of the method for the most part depends on the desired effects.

According to the direct methods the twisters for the fancy yarn making, special knitting machines, etc. are used. According to the indirect methods of manufacture the fancy yarns are produced during initial stages of yarn processing as in mixing chambers, carding, drawing machines, rover. The fibre, sliver or roving so obtained is then processed on spinning machine and/or twisters, that are standard spinning and twisting frames or their modifications with special units. With indirect methods, modifying the conventional processing technologies like dyeing, raising, etc. produces the yarns as well. Some fancy yarns produced according to indirect methods are obtained by weaving and knitting.

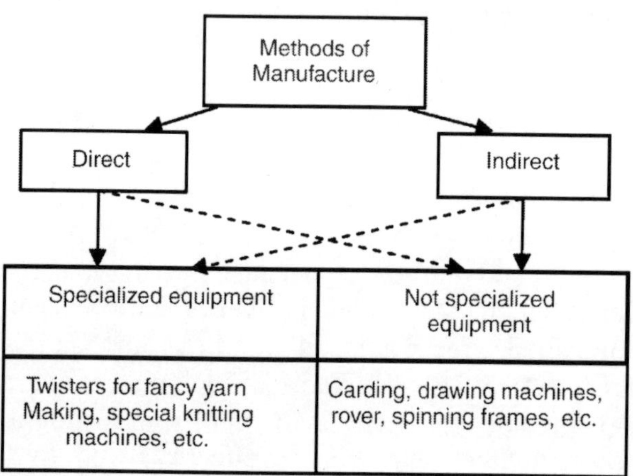

Fig. Diagram to show the Methods of Manufacture of Fancy Yarn.

In the main, it is possible to group the fancy yarns taking into account their use possibilities and raw material as well. Two groups of textiles – the clothing and decorative ones made from woven, knitted fabrics or non-woven materials are the main consumers of fancy yarns. The fancy yarns can be produced in a big variety of raw material combinations. In the case when only a few yarns compose the fancy yarn the grouping according to the raw material can be acceptable.

The number and the type of components virtually change the structure and exterior of the fancy yarn. The number of the components can differ from one to six or more. The components can be not only yarns made from staple fibres or filaments but also the lengths of yarn and/or filament, non-woven material as well as the lever or polymer for the decorative coating of core component.

The discussed criteria: method of manufacture, use of fancy yarn and raw material of components, number and type of components are not suitable for absolutely all cases because some types of the fancy yarn can be manufactured using the different methods but their appearance can be the same. On the other hand, the fancy yarn can be combined from the equal number of the components and the same raw material but their effects can be absolutely different.

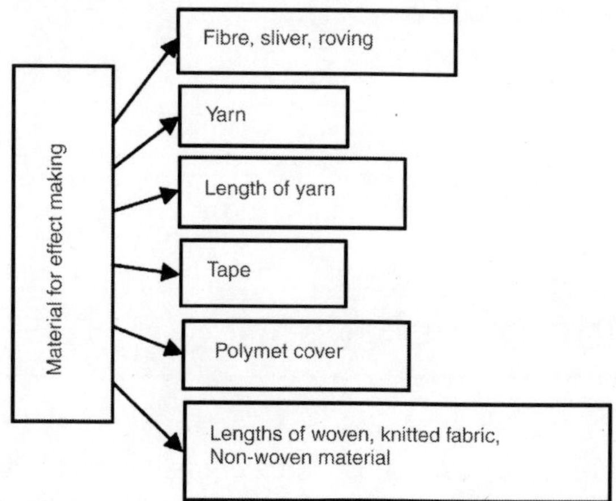

Fig. Grouping of Fancy Yarns According to the Material for Effect Making.

The type of the effect and material for effect making most properly characterise the group of fancy yarn ascription. The grouping according to the type of fancy yarn effect and some examples with colour, structure, and compound effects are demonstrated. The grouping according to the material for effect making.

The quality of such criteria as the type of fancy yarn effect and the material for effect making is that the fancy yarn character here has only one meaning, all assortment of fancy yarns is included-not only traditional fancy yarns but virtually new structure as well, it disappears the necessity to link the effect type and effect making technology.

5

Construction of Felt Fabric

Felt, fabric built up by the interlocking of unspun wool fibres occasionally blended with small quantities of vegetable and synthetic fibres. True felt can only be made of fibres that are covered with minute, flexible, barblike scales, which allow the fibres to interlock when felted. Wool fibres, such as most sheep's fleece, are covered with well-defined scale structures; hair with poor scale definition, such as human hair, is not likely to felt.When felts are intentionally manufactured, fibres and scales are made pliable by heat.

Moisture and, sometimes, soap are used to lubricate as well as soften. The fibres are entangled or matted by agitation such as by rubbing. By the modern method, wool is cleansed, disentangled, and reassembled into thin webs, which are then piled in successive layers to a thickness required for the finished piece.

The prepared fibres are then condensed on a flatbed press. The lower plate of the press is heated as the upper plate oscillates, causing the fibres to interlock. The felt is then fulled, or shrunk, to compress the fibres, which intertwine to form a loose fabric that, after further shrinking, becomes dense and durable.

Fibres used with wool in the manufacture of felt include vegetable fibres, such as cotton, kapok, ramie, and jute, and synthetic fibres, such as rayon and nylon. However, as the percentage of wool is decreased, the propensity to felt is proportionately decreased. In fact, blending wools with other fibres is one means of reducing undesired felting, such as in wool knitted goods that are designed to be machine launderable. Other methods to prevent felting are chemical removal of scales; elimination of the use of tight yarn and fabric constructions, which restrict freedom of fibre movement; and the use of resin coatings, which bind fibres together.

Felt making is one of the primitive arts, antedating weaving.Since the Industrial Revolution, mechanical operations have superseded hand methods, but the principle of felt making has not changed. Finished felts are produced in a range of consistencies from soft cloth to a dense sheet. Roll felts are finished and dyed; sheet felts are sawed, drilled, and turned, like wood, to form polishing wheels and various mechanical parts.

Fig. Felt Machine

Wool felt and other nonwoven fabrics are used in inking pads for automatic printers; in vibration dampeners for machinery; as soundproofing; and for polishing glass, granite, and some metals. Oil-soaked felt pads are used to lubricate machinery. The resilience of felt makes it the only substance suitable for dampers on pianos and other musical instruments. Wool felt is widely used in making hats and other types of clothing.

Woven or knitted wool is felted in a controlled process called fulling or milling. Under extreme conditions, the fabric condenses into a dense, matted structure, making it difficult to identify the original yarns or construction. These felts are often finished by napping or brushing to make them stronger and more resistant to tearing.

Examples of heavily fulled fabrics are loden and melton cloth used for outerwear. Various fabrics of feltlike consistency are distinguished by compound names. The name fur felt identifies fibre origin, and terms such as needle punched, blown, woven, stitched, or quilted felts refer to the manufacturing process. In the manufacture of felt for some applications—roofing felt, papermakers felt—true felting action is not usually employed.

FELT a fabric produced by the "matting" or "felting" together of fibrous materials such as wools, hairs, furs, &c. Most textile fibres possess the quality of matting to some extent, but wools, furs and some few hairs are the only fibres which can be felted satisfactorily. It is probable that the quality of felting must be attributed to the scale structure and waviness of the wools, furs and hairs referred to. When it is desired to incorporate non-felting fibres in felt cloths, wool must be employed to "carry" them.

There are two distinct classes of felts, viz. woven or "threadstructure" felts, and "fibre" or true felts. In the manufacture of thread-structure felts, wools possessing the quality of felting in a high degree are naturally selected, carefully scoured so that the felting quality is not seriously damaged, spun into woollen yarn possessing the necessary fibre arrangement and twist, woven into cloth of such a character that subsequently satisfactory shrinking or felting may be effected, and finally scoured, milled in the stocks of machine of both, dyed and finished on the lines of an ordinary woven fabric. The lighter styles of woven felts may be composed of a single cloth only, but for the heavier styles two or more cloths are woven, one on top of the other, at one and the same time, arrangements being made to stitch the cloths together during the weaving operation.

Fibre felts are exceedingly interesting from the historical point of view. It is now generally admitted that the art of weaving preceded that of spinning, and it must further be conceded that the art of felting preceded that of weaving, so that the felt fabric is probably one of the oldest of the various styles of recognized fabrics.

The inhabitants of the middle and northern regions of Asia seem to have employed felt from time immemorial, as clothing and also as a covering for their habitations. Most of the classical writers refer to it and some of them actually describe its manufacture. Felt was also largely employed by the ancients for their hats, outer garments, and sometimes as a species of armour.

Fibre felts may be divided into three classes, viz. ordinary felts; hat felts; and impregnated felts. As all felts are based upon the ordinary felt, the process of manufacture of this will first be described.

Of the wools employed the principal are: East Indian, German or mid-European, New Zealand crossbreds, and Australian, Cape and Buenos Aires merinos. Vegetable fibres and silk are also employed, but wool must be used to "carry" them; thus a good felting wool may be made to carry its own weight of cotton, hemp, &c. Hairs and furs are principally used in the hat felts.

The average loss upon the wool from the raw state to the finished felt is 40 to 50%. The order of the manufacturing processes is as follows:- mixing, willowing, teasing, scribbling and carding. It is interesting that it is not usual to scour felting wools.

This is not because they are really clean - some are dirty-but because the felting property is liable to be interfered with in the scouring operation. Some wools, however, must be scoured to ensure satisfactory working in the machines. From the card the wool is delivered as a gossamer-like film from 50 to 60 in. wide on to an endless sheet from 30 to 60 yds. long, upon which the felt is built up film upon film until the required thickness - perhaps 4 in is obtained.

To harden this somewhat tender sheet of felt it is now passed through an ironing process, effected by either steamheated rollers - to which a rotatory and vibratory motion is given - playing upon the continually drawn-through cloth; or a huge vibrating flat-iron, to which the cloth is automatically fed, held in position and then wound up while the following length to be treated is drawn under the iron. Soaping, fulling or "felting" and the ordinary finishing operations - including dyeing and printing if desirable - now follow, so that ultimately a strong firm fabric is turned out.

It must be admitted, however, that the strength is much greater lengthwise than cross-wise, owing to the parallelization of the fibres induced in the scribbling and carding operations. Of course, the true felting or contraction occurs in the fulling or felting stock, the fabric being perpetually "hammered" in the presence of fulling agents such as soap, fuller's earth, &c., for a considerable time. The reduction in width, length and thickness is remarkable. This may be controlled within certain limits.

The principal styles of ordinary fibre-felts are - linings for coats, furniture and rubber shoes; saddlery; seatings for carriages and pews; carpets, surrounds and under-felts for carpets; mantles, dresses and table-cloths; felt-slippers; mattress felts; chest-preservers, and shoulderpads; steam-engine packing, motor-car and anti-vibration felts, shipbuilding felts; drawing-roller felts and gun-wad felts.

Hat felts may be divided into two classes, viz. those made from wool and fur respectively. Wool "bodies" used for the lower quality hats are manufactured in the same way as ordinary felts, but the "shape" upon which the film issuing from the carder is built up takes the form of a double cone and thus approximates to the shape of the two hats ultimately formed.

The shape is further controlled and developed in the fulling or felting operation. In the fur hat felts an air-blast is employed to carry the finely separated fibres on to the shape required, upon which shape the fibres are held in position by suction until the required thickness is obtained. The structure is then further developed and "stiffened," *i.e.* impregnated with certain stiffening agents according to requirements.

If desirable the exterior fibres blown on to any shape may be of a different material from the body fabric.

Impregnated felts are simply felts made in the ordinary way but subsequently impregnated with certain agents which give a special quality to the fabric. Messrs McNeill & Co., of London, were the originators of "asphalted-felt" for roofing and, among other styles, place on the market sheathing felt, inodorous felt, dry hair felt, foundation felt, &c., &c. A later development, however, is the impregnated iron-felt manufactured by Messrs Mitchells, Ashworth, Stansfield & Co., of Waterfoot, near Manchester, who not only produce from 70 to 80% of the ordinary felts manufactured in Great Britain, but also place on the market several specialties of which this "iron-felt" is largely used in the construction of bridges, &c., and as a substitute for rubber, it being apparently more durable.

MATTING

MATTING, a general term embracing many coarse woven or plaited fibrous materials used for covering floors or furniture, for hanging as screens, for wrapping up heavy merchandise and for other miscellaneous purposes. In the United Kingdom, under the name of "coir" matting, a large amount of a coarse kind of carpet is made from coco-nut fibre; and the same material, as well as strips of cane, Manila hemp, various grasses and rushes, is largely employed in various forms for making door mats.

Large quantities of the coco-nut fibre are woven in heavy looms, then cut up into various sizes, and finally bound round the edges by a kind of rope made from the same material. The mats may be of one colour only, or they may be made of different colours and in different designs. Sometimes the names of institutions are introduced into the mats. Another type of mat is made exclusively from the above-mentioned rope by arranging alternate layers in sinuous and straight paths, and then stitching the parts together. It is also largely used for the outer covering of ships' fenders. Perforated and otherwise prepared rubber, as well as wire-woven material, are also largely utilized for door and floor mats.

Matting of various kinds is very extensively employed throughout India for floor coverings, the bottoms of bedsteads, fans and fly-flaps, &c.; and a considerable export trade in such manufactures is carried on. The materials used are numerous; but the principal substances are straw, the bulrushes *Typha elephantine* and *T. angustifolia,* leaves of the date palm of the dwarf palm of the Palmyra palm of the coco-nut palm *andof* the screw pine the munja or munj grass and allied grasses, and the mat grasses *Cyperus textilis* and C. *Pangorei,* from the last of which the well-known Palghat mats of the Madras Presidency are made.

Many of these Indian grass-mats are admirable examples of elegant design, and the colours in which they are woven are rich, harmonious and effective in the highest degree. Several useful household articles are made from the different kinds of grasses. The grasses are dyed in all shades and plaited to form attractive designs suitable for the purposes to which they are to be applied. This class of work obtains in India, Japan and other Eastern countries. Vast quantities of coarse matting used for packing furniture, heavy and coarse goods, flax and other plants, &c., are made in Russia from the bast or inner bark of the lime tree.

DRESS FELT

Clothing

Clothing, coverings and garments intended to be worn on the human body. The words *cloth* and *clothing* are related, the first meaning fabric or textile, and the second meaning fabrics used to cover the body. The earliest garments were made of leather and other nonfabrics, rather than of cloth, but these nonfabric garments are included in the category of clothing.

Fashion refers to the kinds of clothing that are in a desirable style at a particular time. At different times in history, fashionable dress has taken very different forms. In modern times nearly everyone follows fashion to some extent. A young woman would look odd if she wore the clothing that her grandmother had worn when young. However, only a small minority of people dress in the clothing that appears in high-fashion magazines or on fashion-show runways.

It is not always easy to tell the difference between basic clothing and fashionable clothing. Especially today, fashion designers often use inexpensive and functional items of clothing as inspiration.

Blue jeans, for instance, originated as functional work clothing for miners and farmers. Yet today, even people who dress in jeans, T-shirts, and sports clothes may be influenced by fashion. One year, fashionable jeans may have narrow legs; the next year the legs may be baggy.

Clothing historians trace the development of dress by studying various sources, including magazines and catalogs, paintings and photographs, and hats, shoes, and other surviving items. Reliable evidence about everyday clothing from the past can be hard to obtain because most publications and images concern the fashions of the wealthy. Furthermore, clothing that has survived from the past tends not to be typical of what was worn in daily life. Museum collections are full of fashionable ball gowns, but have very few everyday dresses worn by ordinary working-class women.

Even fewer examples of ordinary men's clothing have been saved. Such as paintings, prints, and photographs, do provide considerable evidence of the history of everyday clothing. These sources indicate that although everyday clothing does not usually change as rapidly as fashionable dress, it does change constantly.

Fig. Clothing Machine

The Functions of Clothing

Since prehistoric times, people in almost all societies have worn some kind of clothing. Many theories have been advanced as to why humans began to wear clothing. One of the earliest hypotheses is the so-called modesty/shame theory, also known as the fig leaf theory. This theory is based on the biblical story of creation. In the book of Genesis, Adam and Eve, the first human beings, realise they are naked after they eat an apple from the tree of knowledge. Ashamed of their nakedness, they make clothing for themselves out of fig leaves.

As late as the 19th century, most Europeans and Americans believed that people wore clothing primarily for reasons of modesty. With the rise of a nonreligious worldview, however, people began to offer other theories. Some argued that the origin of clothing was functional—to protect the body from the environment. Others argued that some clothing was designed for sexual attraction—to display the body's beauty.

Evidence that early clothing was indeed functional came from a 1991 discovery of a 5,000-year-old male body, frozen on top of a glacier near the Austrian-Italian border. It was clothed in a fur cap, a crudely tanned leather cape, a *loincloth* (strip of cloth wrapped around the waist and between the legs), leggings, and leather shoes. A grass cloak covered the fur and leather clothing.

These clothes would have provided protection against the cold and rain. The Iceman, as he is called, also had tattoos, which may have been marks of decoration or tribal identity, or were perhaps intended to provide magical protection.

Decoration seems to satisfy a fundamental human need. Other animals groom themselves, but only human beings have ornamented themselves. Although in some societies people have worn little or no clothing, so far as we know, people have decorated their bodies in some way in all societies throughout history.

Archaeological and anthropological evidence suggest that early people may have decorated their bodies with paint, tattoos, and other types of ornamentation even before they began wearing clothing made of fur or fabric. Body decoration, like clothing, has served a variety of social and symbolic purposes.

Modern scholars believe that clothing provides a mark of identity and a means of nonverbal communication. In traditional societies, clothing functions almost as a language that can indicate a person's age, gender, marital status, place of origin, religion, social status, or occupation. In modern industrialized societies, clothing is not so rigidly regulated and people have more freedom to choose which messages they wish to convey.

Nevertheless, clothing can still provide considerable information about the wearer, including individual personality, economic standing, even the nature of events attended by the wearer. When a woman who usually wears blue jeans puts on a frilly, flowered dress, she may be stating that she wants to look more traditionally feminine. A person wearing a T-shirt emblazoned with the name of a rock band is probably a fan of that music group and may have attended one of the group's concerts.

A society's economic structure and its culture, or traditions and way of life, also influence the clothing that its people wear. In many societies, religious laws regulated personal behaviour and permitted only members of an elite class to wear certain prestigious items of clothing.

Even in modern democracies, clothing may represent social standing. Clothing with a designer label tends to be relatively expensive, so it may function as an outward sign of a person's economic standing.

Fig. Cloth Combing Machine

Clothing most obviously defines a social role in the case of uniforms, such as those worn by police officers and nurses, and garments worn by clergy or members of religious orders. Clothing metaphors—blue-collar and white-collar workers are used to distinguish between types of work Clothing also derives meaning from the environment in which it is worn. In most cultures brides and grooms as well as wedding guests wear special clothes to celebrate the occasion of a marriage. The clothing worn for rituals such as weddings, graduations, and funerals tends to be formal and governed by unwritten rules that members of the society agree upon.

Clothing may also signal participation in leisure activities. Certain types of recreation, especially active sports, may require specialized clothing. (Example, football, soccer, and hockey players wear matching jerseys and pants designed to accommodate such accessories as protective pads).Most modern societies comprise different social groups, and each group has its own beliefs and behaviours. As a result, different clothing subcultures exist.

Within a single high school, teenagers known as jocks are likely to wear different styles of clothing than teens called nerds. This difference can indicate to which group a teen belongs.

MATERIALS USED FOR CLOTHING

The development of new materials for use as covering or ornament has played a major role in the history of clothing. In early prehistoric times, the range of materials for clothing was small. Until about 10,000 years ago, people used animal skins for clothing. Single skins were worn as capes thrown around the shoulders; two skins fastened together at the shoulder made a simple garment.

Fitted clothes, such as trousers or a *parka* (an outer garment with a hood), were also made from animal skins. Simple needles made out of animal bone, found in many sites in Europe and Asia, provide evidence of sewn leather and fur garments from at least 30,000 years ago. However, only with the development of textile technology did greater variety become possible.

Beginnings of Textile Production

Tens of thousands of years ago, early humans learned to obtain fibres from wild plants, such as wild flax, hemp, and nettles. Such fibres could be spun into thread and made into cloth textiles. People began to weave fabric during the Neolithic Era, a period that began around 8000 BC. Evidence of early weaving comes from fragments of flax fibres found in Switzerland. In some cultures, people made cloth without weaving, by pounding sheets of bark to produce a soft, flexible textile. The development of agriculture led to the domestication of fibre plants, such as cotton, hemp, and flax.

The production of textiles requires the ability to process fibres, spin them into thread, and make cloth from the thread. Cloth can be made in a variety of ways, such as knotting, knitting, and braiding, but most cloth is made by weaving. Weaving is usually done on a loom that holds long threads (called the warp) under uniform tension so that other threads (the woof or weft) can be inserted over and under them. Many different types and patterns of weaving are possible, depending on the fibre used and the arrangement of the threads. Ancient Egyptians wove their earliest textiles from flax, which produced linen; in southern Europe, the earliest textiles came from wool; in China, from silk; and in India, Peru, and Cambodia, from cotton.

Wool

Wool was probably the first animal fibre to be made into cloth. People began to raise sheep for wool about 6,000 years ago. Different kinds of wool, and mixtures of wool with other fibres, can be used to create tightly woven fabrics with smooth surfaces or more loosely woven fabrics with rougher surfaces. In many Middle Eastern societies, nomadic peoples beat wool into a matted fabric called felt, which they used to make clothing, saddle blankets, tents, and other useful items. Sheep-raising nomads in the Middle East also invented carpets made by knotting woolen weft threads onto a linen warp.

Woolens came to be the characteristic textile for European clothing during the Middle Ages, which lasted from about the 5th to the 15th century. People still commonly use wool for clothing, either as pure wool textiles or in blends with other fibres. Wool has obvious advantages as a clothing fabric because it is warm, even when wet, and readily accepts a wide range of colour dyes.

Silk

China's chief contribution to world clothing has been the development of silk thread and cloth. By 3000 BC the Chinese had domesticated silkworms, feeding them mulberry leaves and unwinding their fresh cocoons to produce long strands of silk fibre. This fibre was spun into thread, and the thread was woven into cloth. By 1500 BC elaborate weaving techniques had been developed, using thread dyed in many colours. Silk is lustrous, soft, and lightweight but warm, and it can easily be dyed. Silk cloth excavated from tombs dating from the 2nd century BC includes *gauze* (thin, loosely woven fabric), *twill* (fabric with a woven design of parallel diagonal ribs), *damask* (fabric woven with patterns on both sides), *brocade* (heavy fabric woven with an intricate raised design), and plain cloth embroidered with different stitches. Farm women in China of the period were expected to raise silkworms and produce silk as part of their regular household duties. Silk was used in China, and it was also exported along the Silk Road; this ancient trade route linking China and the Roman Empire was named after the primary export carried on it. The silk trade, conducted between western Asia and the Mediterranean as early as AD 200, brought great wealth to ancient China and sustained the economies of towns along the route. China kept the technology of silk production secret; the ancient Greeks speculated that silk grew on a special tree in China. Christian monks finally broke China's monopoly on silk production in the 400s, when they smuggled silkworm eggs to Syria on their return from China.

Cotton

Domesticated cotton first came into widespread use in ancient India around 3000 BC. Much Indian cotton cloth had a simple weave to take advantage of the material's lightness and airiness; it was then dyed and printed by hand using wood blocks that had patterns cut into them. Indian textiles have influenced textiles of other regions from antiquity through the modern era.

One type of printed cotton fabric, called calico by Europeans (after an Indian textile centre named Calicut), was exported from India to Europe in large quantities in the 16th century. Imported Indian cotton became extremely fashionable and undermined the European woolen cloth industry. This helped stimulate the Industrial Revolution of the 18th century, as Europeans tried to develop ways of producing cotton cloth by machine more cheaply than it could be made by hand in India.

REGIONAL DIFFERENCES IN TEXTILES

The materials used for clothing vary around the world. Some textiles are better suited to a particular climate. Knitted wool is more useful in cold climates, and thin woven cotton is more useful in warm climates. For most of history, the textiles people used depended on the raw materials available locally, such as flax in Egypt, cotton in India, and silk in China. Beyond considerations of utility and availability, however, people tend to derive regional or national identity from their most characteristic textiles, just as they do from their typical foods.

Thus, a European businessman defines himself in part by his woolen suit, an Indonesian farmer by his cotton *sarong* (skirt of brightly coloured cloth, worn wrapped about the waist). For centuries, silk-wearing Chinese people despised wool, which they considered the fabric of uncivilized people. Today, such considerations of identity have weakened amid international trade and international cultural exchange.

RECENT TRENDS IN TEXTILE MANUFACTURE

Throughout the 20th century, the trend in textiles has been toward lighter-weight materials. Better transportation and improved indoor heating have made warmth a less important aspect of clothing for most people than it was in earlier centuries. Since 1900 the weight of average clothing fabric has fallen by more than one-third. When people need extremely warm clothes, they tend to wear special outer garments such as parkas that today are typically made of lightweight synthetic materials.

Synthetic materials, produced from chemical compounds rather than from plant fibres or animal hair, have provided less-expensive substitutes for natural fabrics. Synthetic materials can also be superior to natural fibres in strength and durability.

Rayon was an early synthetic substitute for silk. Nylon, a synthetic fabric introduced in the 1930s, was another early substitute for silk and quickly became the fibre of choice for women's stockings. Polyester, a form of plastic, was introduced in clothing in the early 1950s. Blended with rayon or cotton, polyester found its first use in so-called wash-and-wear fabrics that needed little or no ironing.

Synthetic fibres fell out of favour in the late 1960s but new kinds of polyester that are more durable and have a softer, more natural feel to them have become increasingly popular in the late 20th century. Synthetic fibres such as spandex have revolutionized clothing by making possible the production of extremely flexible, form-fitting garments. Other synthetic fibres, created for special purposes, range from lightweight but extremely warm or water-resistant fabrics, such as polypropylene and the composite polymer Gore-tex, to woven, bullet-proof fabrics such as Kevlar that serve as body armor.

HAT FELT TECHNOLOGY

HAT, a covering for the head worn by both sexes, and distinguished from the cap or bonnet by the possession of a brim. The word in O.E. is *hcet,* which is cognate with O. Frisian *hatt,* O.N. *hotte,* &c., meaning head-covering, hood; it is distantly related to the hood, which is cognate with the German for "hat," *Hut.* The history of the hat as part of the apparel of both sexes, with the various changes in shape which it has undergone, is treated in the article Costume.

Hats were originally made by the process of felting, and as tradition ascribed the discovery of that very ancient operation to St Clement, he was assumed as the patron saint of the craft. At the present day the trade is divided into two distinct classes.

The first and most ancient is concerned with the manufacture of felt hats, and the second has to do with the recent but now most extensive and important manufacture of silk or dress hats. In addition to these there is the important manufacture of straw or plaited hats and hats are occasionally manufactured of materials and by processes not included under any of these heads, but such manufactures do not take a large or permanent position in the industry.

Felt Hats: There is a great range in the quality of felt hats: the finer and more expensive qualities are made entirely of fur; for commoner qualities a mixture of fur and wool is used; and for the cheapest kinds wool alone is employed. The processes and apparatus necessary for making hats of fur differ also from those required in the case of woollen bodies; and in large manufactories machinery is now generally employed for operations which at no distant date were entirely manual.

An outline of the operations by which the old beaver hat was made will give an idea of the manual processes in making a fur napped hat, and the apparatus and mechanical processes employed in making ordinary hard and soft felts will afterwards be noticed.

Hatters' fur consists principally of the hair of rabbits (technically called coneys) and hares, with some proportion of nutria, musquash and beavers' hair; and generally any parings and cuttings from furriers are also used. Furs intended for felting are deprived of their long coarse hairs, after which they are treated with a solution of nitrate of mercury, an operation called carroting or *secretage,* whereby the felting properties of the fur are greatly increased. The fur is then cut by hand or machine from the ski and in this state it is delivered to the hat maker.

The old process of making a beaver hat was as follows. The materials of a proper beaver consisted, for the body or foundation, of rabbits' fur, and for the nap, of beaver fur, although the beaver was often mixed with or supplanted by a more common fur.

In preparing the fur plate, the hatter weighed out a sufficient quantity of rabbit fur for a single hat, and spread it out and combined it by the operation of bowing. The bow or stang ABC was about 7 ft. long, and it stretched a single cord of catgut which the workman vibrated by means of a wooden pin E, furnished with a half knob at each end. Holding the bow in his left hand, and the pin in his right, he caused the vibrating string to come in contact with the heap of tangled fur, which did not cover a space greater than that of the hand.

At each vibration some of the filaments started up to the height of a few inches, and fell away from the mass, a little to the right of the bow, their excursions being restrained by a concave frame of wicker work called the basket. One half of the material was first operated on, and by bowing and gathering, or a patting use of the basket, the stuff was loosely matted, about 50 by 36 in., called a bat. In this formation care was taken to work about two-thirds of the fur down towards what was intended for the brim, and this having been effected, greater density was induced by gentle pressure with the basket.

It was then covered with a wettish linen cloth, upon which was laid the hardening skin, a piece of dry half-tanned horse hide. On this the workman pressed until the stuff adhered closely to the damp cloth, in which it was then doubled up, freely pressed with the hand, and laid aside. By this process, called basoning, the bat became compactly felted and thinned toward the sides and point.

The other half of the fur was next subjected to precisely the same processes, after which a cone-shaped slip of stiff paper was laid on its surface, and the sides of the bat were folded over its edges to its form and size. It was then laid paper-side downward upon the first bat, which was now replaced on the hurdle, and its edges were transversely doubled over the introverted side-lays of the second bat, thus giving equal thickness to the whole body. In this condition it was reintroduced between folds of damp linen cloth, and again hardened, so as to unite the two halves, the knitting together of which was quickly effected. The paper was then withdrawn, and the body in the form of a large cone removed to the plank or battery room. The battery consisted of an open iron boiler or kettle A filled with scalding hot water, with shelves, B, C, partly of mahogany and partly of lead, slop ing down to it. Here the body was first dipped in the water, and then withdrawn to the plank to cool and drain, when it was unfolded, rolled gently with a pin tapering towards the ends, turned, and worked in every direction, to toughen and shrink it, and at the same time prevent adhesion of its sides.

Stopping or thickening any thin spots seen on looking through the body, was carefully performed by dabbing on additional stuff in succes sive supplies from the hot liquor with a brush frequently dipped into the kettle, until the body was shrunk sufficiently (about one-half) and thoroughly equalized. When quite dried, stiffening was effected with a brush dipped into a thin varnish of shellac, and rubbed into the body, the surface intended for the inside having much more laid on it than the outer, while the brim was made to absorb many times the quantity applied to any other part. On being again dried, the body was ready to be covered with a nap of beaver hair. For this, in inferior qualities, the hair of the otter, nutria or other fine fur was sometimes substituted. The requisite quantity of one or other of these was taken and mixed with a proportion of cotton, and the whole was bowed up into a thin uniform lap. The cotton merely served to give sufficient body to the material to enable the workman to handle the lap.

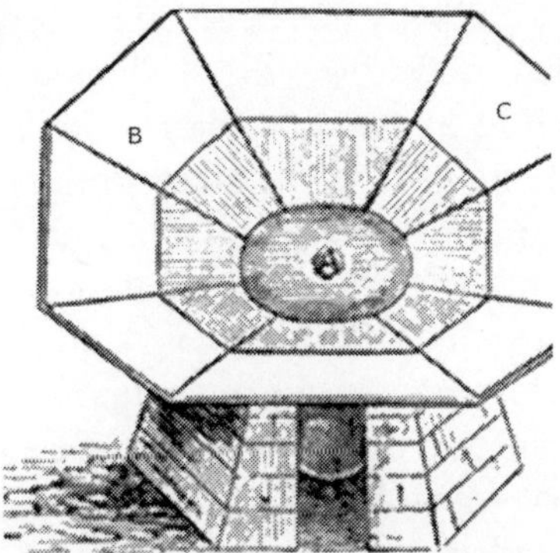

Fig. Hat Felt Machine

Being damped, the workman spread over it a covering of this lap, and by moistening and gentle patting with a brush the cut ends of the hair penetrated and fixed themselves in the felt body.

The hat was then put into a coarse hair cloth, dipped and rolled in the hot liquor until the fur was quite worked in, the cotton being left on the surface loose and ready for removal. The blocking, dyeing and finishing processes in the case of beaver hats were similar to those employed for ordinary felts, except that greater care and dexterity were required on the part of the workmen, and further that the coarse hairs or kemps which might be in the fur were cut off by shaving the surface with a razor. The nap also had to be laid in one direction, smoothed and rendered glossy by repeated wettings, ironings and brushings. A hat so finished was very durable and much more light, cool and easy-fitting to the head than the silk hat which has now so largely superseded it.

The first efficient machinery for making felt hats was devised from the United States the machine-making processes were introduced into England about the year 1858; and now in all large establishments machinery such as that alluded to below is employed. For the forming of hat bodies two kinds of machine are used, according as the material employed is fur or wool. In the case of fur, the essential portion of the apparatus is a "former," consisting of a metal cone of the size and form of the body or bat to be made, perforated all over with small holes. The cone is made to revolve on its axis slowly over an orifice under which there is a powerful fan, which maintains a strong inward draught of air through the holes in the cone.

At the side of the cone, and with an opening towards it, is a trunk or box from which the fur to be made into a hat is thrown out by the rapid revolution of a brushlike cylinder, and as the cloud of separate hairs is expelled from the trunk, the current of air being sucked through the cone carries the fibres to it and causes them to cling closely to its surface. Thus a coating of loose fibres is accumulated on the copper cone, and these are kept in position only by the exhaust at work under it.

When sufficient for a hat body has been deposited, it is damped and a cloth is wrapped round it; then an outer cone is slipped over it and the whole is removed for felting, while another copper cone is placed in position for continuing the work. The fur is next felted by being rolled and pressed, these operations being performed partly by hand and partly by machine.

In the case of wool hats the hat or body is prepared by first carding in a modified form of carding machine. The wool is divided into two separate slivers as delivered from the cards, and these are wound simultaneously on a double conical block of wood mounted and geared to revolve slowly with a reciprocating horizontal motion, so that there is a continual crossing and recrossing of the wool as the sliver is wound around the cone.

This diagonal arrangement of the sliver is an essential feature in the apparatus, as thereby the strength of the finished felt is made equal in every direction; and when strained in the blocking the texture yields in a uniform manner without rupture. The wool wound on the double block forms the material of two hats, which are separated by cutting around the median or base line, and slipping each half off at its own end. Into each cone of wool or bat an "inlayer" is now placed to prevent the inside from matting, after which they are folded in cloths, and placed over a perforated iron plate through which steam is blown.

When well moistened and heated, they are placed between boards, and subjected to a rubbing action sufficient to harden them for bearing the subsequent strong planking or felting operations. The planking of wool hats is generally done by machine, in some cases a form of fulling mill being used; but in all forms the agencies are heat, moisture, pressure, rubbing and turning.

When by thorough felting the hat bodies of any kind have been reduced to dense leathery cones about one-half the size of the original bat, they are dried, and, if hard felts are to be made, the bodies are at this stage hardened or stiffened with a varnish of shellac. Next follows the operations of blocking, in which the felt for the first time assumes approximately the form it is ultimately to possess.

For this purpose the conical body is softened in boiling water, and forcibly drawn over and over a hat-shaped wooden block.

The operation of dyeing next follows, and the finishing processes include shaping on a block, over which crown and brim receive ultimately their accurate form, and pouncing or pumicing, which consists of smoothing the surface with fine emery paper, the hat being for this purpose mounted on a rapidly revolving block. The trimmer finally binds the outer brim and inserts the lining, after which the brim may be given more or less of a curl or turn over according to prevailing fashion.

Silk Hats

The silk hat, which has now become co-extensive with civilization, is an article of comparatively recent introduction. It was invented in Florence about 1760, but it was more than half a century before it was worn to any great extent.

A silk hat consists of a light stiff body covered with a plush of silk, the manufacture of which in a brilliant glossy condition is the most important element in the industry. Originally the bodies were made of felt and various other materials, but now calico is chiefly used. The calico is first stiffened with a varnish of shellac, and then cut into pieces sufficient for crown, side and brim. The side-piece is wound round a wooden hat block, and its edges are joined by hot ironing, and the crown-piece is put on and similarly attached to the side. The brim, consisting of three thicknesses of calico cemented together, is now slipped over and brought to its position, and thereafter a second side-piece and another crown are cemented on.

The whole of the body, thus prepared, now receives a coat of size, and subsequently it is varnished over, and thus it is ready for the operation of covering. In covering this body, the under brim, generally of merino, is first attached, then the upper brim, and lastly the crown and side sewn together are drawn over.

All these by hot ironing and stretching are drawn smooth and tight, and as the varnish of the body softens with the heat, body and cover adhere all over to each other without wrinkle or pucker. Dressing and polishing by means of damping, brushing and ironing, come next, after which the hat is "velured" in a revolving machine by the application of haircloth and velvet velures, which cleans the nap and gives it a smooth and glossy surface. The brim has only then to be bound, the linings inserted, and the brim finally curled, when the hat is ready for use.

RUGS AND CARPETS

Rugs and Carpets, heavy fabrics commonly made of wool and used as floor coverings. A rug differs from a carpet in that it is usually woven in one piece and can be of any size but usually does not cover an entire floor.The term *rug* is derived from the Scandinavian word *rugga* by way of the old Norwegian word *rogg,* which meant a wool covering for the bed or body. For several centuries in Europe, the term *rug* denoted a rough, heavy woolen fabric characterized by a coarse, napped finish and used as apparel by the poorer classes.

The term *carpet* was used originally to describe coverings for tables, beds, and other furniture, and only from the early 18th century was it associated with the floor. The modern usage is imprecise and includes all woven floor coverings and some textiles, such as wall hangings, furniture coverings, and saddlebags, made with a knotted pile or woven like a tapestry. The word *carpet* is ultimately derived from the Latin *carpere,* "to pluck or seize," thus implying a plucking of wool or carding of wool fibres, and reflects the fact that for centuries wool has been used in making carpets.

FLAT-SURFACED WEAVES

The earliest carpets, made before the first millennium BC, were flat-surfaced weaves.

Tapestry

Tapestry is probably the oldest of the flat-surfaced patterned carpet weaves. It is characterized by wefts, or horizontal threads, that do not run the full width of the fabric. Instead, discontinuous wefts of different colours form the design patterns.

Tapestry looms vary in form from culture to culture, although in general they have two stationary parallel bars and one or two adjustable bars around which the warp (vertical threads) is secured. They may be either vertical or horizontal, with heavy beams between which the warp is stretched. Auxiliary bars are attached by means of loops, or heddles, to alternate warps and enable the weaver to pass weft threads between odd and even warps. The warp-weighted loom, a type of upright tapestry loom, was used in ancient Greece and remains in use in Scandinavia. It is so named because the warp hangs free from the upper bar, weighted at the ground with ceramic or stone weights tied to the end of the warp.

Soumak

Soumak, another technique for making flat-woven rugs, originated in the Middle East as early as the 7th century BC; fine contemporary specimens are made in Caucasia. Wefts are wrapped onto the warps in a weft-wise (lateral) direction. Frequently, rows of soumak are alternated with rows of plain weave.

Others

Other flat-woven rug techniques include brocading (introducing supplementary wefts or warps to create a pattern) and embroidery. In some cases, two or more techniques may be used in the production of a single rug.

KNOTTED-PILE WEAVES

The characteristic feature of pile carpets is the knotting and fastening of extra strands of material into a foundation weave to form a dense layer of loops that completely covers the foundation on one side.

Short individual strands are manually wrapped or knotted around taut warp threads. After each row is knotted, a weft thread is carried across the full width of the web and beaten firmly into place. The warp runs the length of the carpet, and the fineness of the weave depends on the thickness of the warps and their proximity to one another. Generally, looms for weaving knotted-pile carpets are the same as those for tapestry.

MATERIALS

The most common material for rugs and carpets has traditionally been sheep's wool, although in certain regions goat's or camel's hair is also used. Luxury carpets are woven with silk pile. Although wool is often used for warp, cotton is more common because of its smooth surface and its resistance to stretching. When gold and silver threads are used, they are generally brocaded into the woven surface.

Natural Dyes

Until the late 19th century, natural dyes were used for colouring weaving yarns. Information about dyeing practices before the 16th century is limited, but it is known that all dye matter came from mineral pigments, from vegetal (plant) materials, or from animals and insects.

Mineral pigments such as ocher (yellow or red), limestone or lime (white), manganese (black), cinnabar (red), blue azurite, green malachite, lead oxide (red), and lapis lazuli (blue) were probably the first materials used for dyeing. Vegetal dyes came from leaves, roots, bark, and, occasionally, fruit or flowers of plants. Woad, a plant of the mustard family, and indigo, a bush from the pea family, were both used for blue dye.

Four vegetal yellows were especially important—saffron, safflower, weld, and fustic. Madder and redwoods had been used since ancient times for orange colours, and browns and blacks came from logwood, betel nuts, walnut and butternut hulls, and a gum resin called cutch.

Animal sources of dyes included shield scales or cochineal insects found on cacti in North and South America, and kermes, a scale insect found on oak trees near the Mediterranean; both produced reds and pinks.

6

Techniques of Fabric Contruction

INTRODUCTION

Fabric is any piece of cloth.

Fabrics can be made or constructed by using a number of techniques as given below:

- *Weaving*: By interlacing of yarns
- *Knitting*: By interlocking the loops
- *Felting*: From matting of the wool fibres
- *Bonding*: By adhering the fibres to each other
- *Knotting*: By tying the fabrics.

WEAVING

In weaving two sets of yarns are interlaced at right angle to one another in an established sequence. *e.g*. 'charpai' being made? On the frame, first one rope is taken and interlooped on two parallel edges and after that a second rope is interlaced perpendicular to it going once over the rope and once under it. This kind of an interlacement gives an even check effect and the weave is quite firm. The weaving of a fabric is also done in a similar way, except for the fact that yarns are used for interlacing and a loom is used to hold the thread instead of a frame.

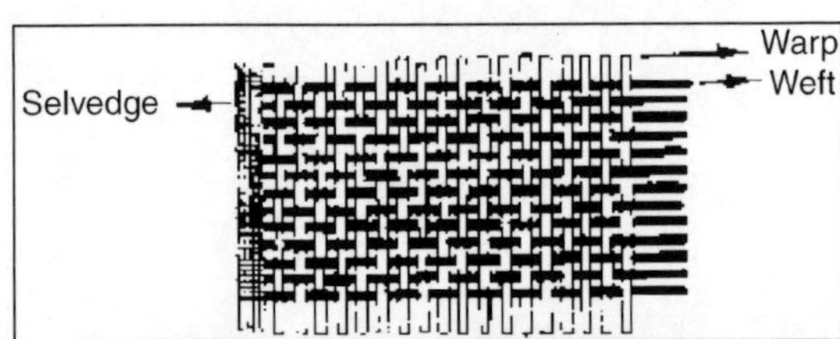

Fig. Common Terms in Textiles

- *Selvedge*: When you examine a fabric, you see two long finished edges, one on each side along the length of the fabric. These edges are called selvedge and give strength to the edges which is important in further processing of the fabric.

- *Warp*: They are the yarns along the length of the fabrics or parallel to the selvedge. They are also called ends.
- *Weft*: Woven fabric, besides warps, another set of yarns move perpendicular to warps. They are called wefts. They are interlaced with warps in a crosswise direction to make a fabric. They are also called picks or fillings.
- *Thread Count*: Some woven fabrics look dense and compact whereas others open. This difference is due to the thread count which refers to the total number of warps and wefts per square inch of a woven fabric. Thread count tells us about the fabric quality and durability. Fabric with a higher thread count is better than fabric with lower thread count. Also, for good quality fabric, warps and wefts should be more or less equal in number.
- Weaving looms can range from quite simple to very complex. Looms have been used to produce cloth for thousands of years, and while technology has improved the loom, the basic strategies and practices remain much the same. Understanding the parts of the loom can help you learn to weave or simply learn a bit more about how weaving works.
- *Warp Beam*: The warp beam is a beam that holds the warp, or vertical threads, on the loom. Depending on the loom, this can be a simple beam or a more complex arrangement allowing for easier warping of the loom.
- *Heddles*: Heddles are made of cord or wire, and are attached to the shaft of the loom. The warp threads pass through the heddles, separating the warp threads to allow the weft threads to pass between them easily.
- *Harnesses*: The more harnesses or shafts a loom has, the more design possibilities you will have. Most larger looms have four harnesses, and table looms often only one; however, looms of up to 16 harnesses are available.

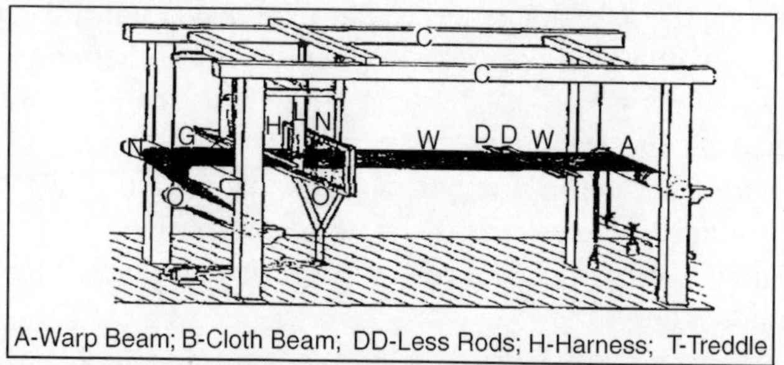

A-Warp Beam; B-Cloth Beam; DD-Less Rods; H-Harness; T-Treddle

- *Shuttle*: The shuttle is a bobbin which holds the weft yarns. The shuttle is passed or thrown back and forth to create the weft of the fabric.

- *Beater and Reed*: The beater of the loom presses the newly created weft thread against the already woven fabric using a reed. This process is called battening.
- *Cloth Beam or Cloth Roll*: It is located at the end of the loom on which the fabric is rolled as it is constructed. It also supports the warp yarn at the front side of the loom.

STEPS IN WEAVING PROCESS

Weaving process is completed in four steps:

1. *Shedding*: Raising the required wrap yarns by using the harness and forming the shed.
2. *Picking*: Inserting the shuttle containing filling yarns through the shed.
3. *Battening*: Pushing the filling yarns in place with the help of reed to form the compact structure. Its also called beating up.
4. *Taking Up and letting Off*: Winding the finished fabric on the cloth beam and releasing more of the warp from the warp beam.

CONSTRUCTION OF A PLAIN WEAVE

Plain weave is the simplest interlacing pattern which can be produced. It is formed by alternatively lifting and lowering one warp thread across one weft thread. A plain weave fabric in plan view and warp way and weft way cross-sections through the same fabric. The diagrams are idealized because yarns are seldom perfectly regular and the pressure between the ends and picks tends to distort the shape of the yarn cross-sections unless the fabrics are woven from monofilament yarns or strips of film. The yarns also do not lie straight in the fabric because the warp and weft have to bend round each other when they are interlaced. The wave form assumed by the yarn is called crimp and is referred to in greater.

Diversity of Plain Weave Fabrics

The characteristics of the cloths woven will depend on the type of fibre used for producing the yarn and whether it is a monofilament yarn, a flat, twisted or textured (multi-)filament yarn or whether it has been spun from natural or manufactured staple fibres. The stiffness of the fabrics and its weavability will also be affected by the stiffness of the raw materials used and by the twist factor of the yarn, that is the number of turns inserted in relation to its linear density. Very highly twisted yarns are sometimes used to produce special features in plain weave yarns. The resulting fabrics may have high extensibility or can be semiopaque.

The area density of the fabric can be varied by changing the linear density or count of the yarns used and by altering the thread spacing, which affects the area covered by the yarns in relation to the total area. The relation between the thread spacing and the yarn linear density is called the cover factor. Changing the area density and/or the cover factors may affect the strength, thickness, stiffness, stability, porosity, filtering quality and abrasion resistance of fabrics.

Square sett plain fabrics, that is fabrics with roughly the same number of ends and picks per unit area and warp and weft yarns of the same linear densities are produced in the whole range of cloth area densities and cloth cover factors. Low area density fabrics of open construction include bandages and cheese cloths, light area density high cover factor fabrics include typewriter ribbons and medical filter fabrics, heavy open cloths include geotextile stabilization fabrics and heavy closely woven fabrics include cotton awnings.

Warp-faced plain fabrics generally have a much higher warp cover factor than weft cover factor. If warp and weft yarns of similar linear density are used, a typical warp faced plain may have twice as many warp ends as picks. In such fabrics the warp crimp will be high and the weft crimp extremely low. The plan view and cross-sections of such a fabric. By the use of suitable cover factors and choice of yarns most of the abrasion on such a fabric can be concentrated on the warp yarns and the weft will be protected.

Weft-faced plains are produced by using much higher weft cover factors than warp cover factors and will have higher weft than warp crimp. Because of the difference in weaving tension the crimp difference will be slightly lower than in warp-faced plain fabrics. Weft-faced plains are little used because they are more difficult and expensive to weave.

Construct a Point Paper Diagram

To illustrate a weave either in plan view and/or in cross-section, takes a lot of time, especially for more complicated weaves. A type of shorthand for depicting weave structures has therefore been evolved and the paper used for producing designs is referred to as squared paper, design paper or point paper. Generally the spaces between two vertical lines represent one warp end and the spaces between two horizontal lines one pick. If a square is filled in it represents an end passing over a pick whilst a blank square represents a pick passing over an end. If ends and picks have to be numbered to make it easier to describe the weave, ends are counted from left to right and picks from the bottom of the point paper design to the top.

The point paper design is the design for a plain weave fabric. To get a better impression of how a number of repeats would look, four repeats of a design (two vertically and two horizontally) are sometimes shown. When four repeats are shown the first repeat is drawn in the standard way but for the remaining three repeats crossing diagonal lines may be placed into the squares, which in the firsts repeat, are filled in. This method is shown for a plain weave.

Plan View

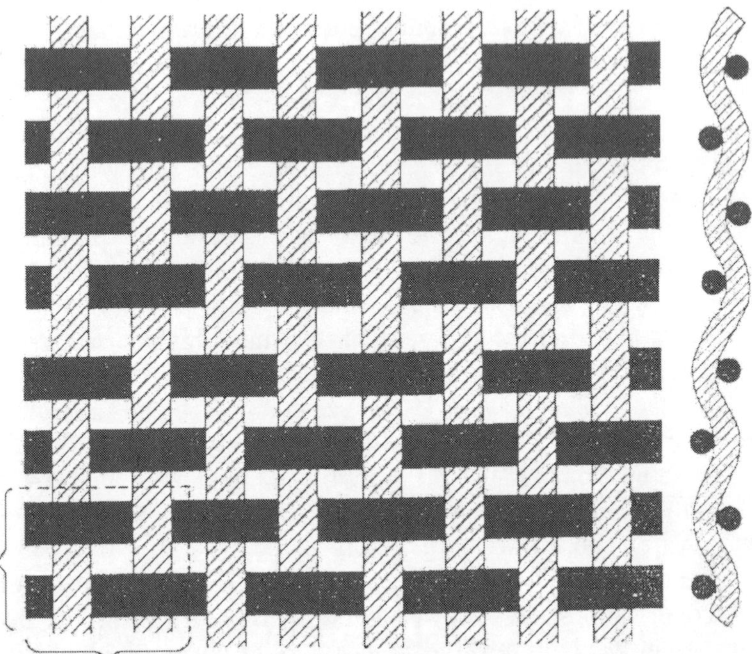

Fig. Fabric woven with plain weave - plan view (4x4 repeats) and warp and weft cross-sections.

RIB FABRICS AND MATT WEAVE FABRICS

These are the simplest modifications of plain weave fabrics. They are produced by lifting two or more adjoining warp threads and/or two or more adjoining picks at the same time. It results in larger warp and/or weft covered surface areas than in a plain weave fabric. As there are fewer yarn intersections it is possible to insert more threads into a given space, that is to obtain a higher cover factor, without jamming the weave.

Matt Fabrics (or hopsack)

Simple matt (or hopsack) fabrics have a similar appearance to plain weave. The simplest of the matt weaves is a 2/2 matt, where two warp ends are lifted over two picks, in other words it is like a plain weave fabric with two ends and two picks weaving in parallel. The number of threads lifting alike can be increased to obtain 3/3 or 4/4 matt structures. Special matt weaves, like a 4/2 matt are produced to obtain special technical effects. Larger matt structures give the appearance of squares but are little used because they tend to become unstable, with long floats and threads in either direction riding over each other. If large matt weaves are wanted to obtain a special effect or appearance they can be stabilized by using fancy matt weaves containing a binding or stitching lift securing a proportion of the floats.

Matt weave fabrics can be woven with higher cover factors and have fewer intersections. In close constructions they may have better abrasion and better filtration properties and greater resistance to water penetration. In more open constructions matt fabrics have a greater tear resistance and bursting strength. Weaving costs may also be reduced if two or more picks can be inserted at the same time.

Rib Fabrics

In warp rib fabrics there are generally more ends than picks per unit length with a high warp crimp and a low weft crimp and vice versa for weft rib fabrics. The simplest rib fabrics are the 2/2 warp rib and the 2/2 weft rib shown respectively. In the 2/2 warp rib one warp end passes over two picks whilst in the 2/2 weft rib one pick passes over two adjoining ends. The length of the floats can be extended to create 4/4,6/6,3/1 or any similar combination in either the warp or weft direction. 3/1 and 4/4 warp ribs.

In rib weaves with long floats it is often difficult to prevent adjoining yarns from overlapping. Weft ribs also tend to be expensive to weave because of their relatively high picks per unit length which reduces the production of the weaving machine unless two picks can be inserted at the same time.

Twill Fabrics

A twill is a weave that repeats on three or more ends and picks and produces diagonal lines on the face of a fabric. Such lines generally run from selvedge to selvedge. The direction of the diagonal lines on the surface of the cloth are generally described as a fabric is viewed along the warp direction. When the diagonal lines are running upwards to the right they are Z twill' or 'twill right' and when they run in the opposite direction they are 'S twill' or 'twill left'. Their angle and definition can be varied by changing the thread spacing and/or the linear density of the warp and weft yarns. For any construction twills will have longer floats, fewer intersections and a more open construction than a plain weave fabric with the same cloth particulars.

Industrial uses of twill fabrics are mainly restricted to simple twills and only simple twills are discussed here. Broken twills, waved twills, herringbone twills and elongated twills are extensively used for suiting and dress fabrics. For details of such twills see Robinson and Marks or Watson. The smallest repeat of a twill weave consists of 3 ends x 3 picks. There is no theoretical upper limit to the size of twill weaves but the need to produce stable fabrics with floats of reasonable length imposes practical limits. The twill is produced by commencing the lift sequence of adjacent ends on one pick higher or one pick lower. The lift is the number of picks which an end passes over and under.

In a 2/1 twill, an end will pass over two picks and under one, whilst in a 1/2 twill the end will pass over one pick and then under two picks.

Either weave can be produced as a Z or S twill. There are, therefore, four combinations of this simplest of all twills. To show how pronounced the twill line is even in a 3 x 3 twill, four repeats (2 x 2) lifted ends shown in solid on the point paper. 2/1 twills are warp faced twills, that is fabrics where most of the warp yarn is on the surface, whilst 1/2 twills have a weft face. Weft-faced twills impose less strain on the weaving machine than warp-faced twills because fewer ends have to be lifted to allow picks to pass under them. For this reason, warp-faced twills are sometimes woven upside down, that is as weft-faced twills. The disadvantage of weaving twills upside down is that it is more difficult to inspect the warp yarns during weaving.

Lenos

In lenos adjoining warp ends do not remain parallel when they are interlaced with the weft but are crossed over each other. In the simplest leno one standard end and one crossing end are passed across each other during consecutive picks. Two variations of this structure are shown in plan view and cross-section. Whenever the warp threads cross over each other, with the weft passing between them, they lock the weft into position and prevent weft movement.

Leno weaves are therefore used in very open structures, such as gauzes, to prevent thread movement and fabric distortion. When the selvedge construction of a fabric does not bind its edge threads into position, leno ends are used to prevent the warp threads at each side of a length of cloth from slipping out of the body of the fabric. They are also used in the body of fabrics when empty dents are left in weaving because the fabrics are to be slit into narrower widths at a later stage of processing.

Leno and gauze fabrics may consist of standard and crossing ends only or pairs or multiples of such threads may be introduced according to pattern to obtain the required design. For larger effects standard and crossing ends may also be in pairs or groups of three. Two or more weft threads may be introduced into one shed and areas of plain fabric may be woven in the weft direction between picks where warp ends are crossed over to give the leno effect. Gauze fabrics used for filtration generally use simple leno weaves.

Standard and crossing ends frequently come from separate warp beams. If both the standard and crossing ends are warped on to one beam the same length of warp is available for both and they will have to do the same amount of bending, that is they will have the same crimp. Such a leno fabric is shown in plan view and cross-section. If the two series of ends are brought from separate beams the standard ends and the crossing ends can be tensioned differently and their crimp can be adjusted separately. In such a case it is possible for the standard ends to lie straight and the crossing ends to do all the bending. That crossing threads can be moved either from the right to the left or from the left to the right on the same pick and adjoining leno pairs may either cross in the same or opposite directions. The direction of crossing can affect locking, especially with smooth monofilament yarns. When using two beams it is also possible to use different types or counts of yarn for standard and crossing ends for design or technical applications. The actual method of weaving the leno with doups or similar mechanisms is not discussed in this chapter.

When lenos are used for selvedge construction only one to four pairs of threads are generally used at each side and the leno selvedge is produced by a special leno mechanism that is independent of the shedding mechanism of the loom. The leno threads required for the selvedge then come from cones in a small separate creel rather than from the warp beam. The choice of selvedge yarns and tensions is particularly important to prevent tight or curly selvedges. The crimps selected have to take into account the cloth shrinkage in finishing.

Satins and Sateens

In Britain a satin is a warp-faced weave in which the binding places are arranged to produce a smooth fabric surface free from twill lines. Satins normally have a much greater number of ends than picks per centimetre. To avoid confusion a satin is frequently described as a 'warp satin'. A sateen, frequently referred to as a 'weft sateen', is a weft-faced weave similar to a satin with binding places arranged to produce a smooth fabric free of twill lines.

Sateens are generally woven with a much higher number of picks than ends. Satins tend to be more popular than sateens because it is cheaper to weave a cloth with a lower number of picks than ends. Warp satins may be woven upside down, that is as a sateen but with a satin construction, to reduce the tension on the harness mechanism that has to lift the warp ends.

To avoid twill lines, satins and sateens have to be constructed in a systematic manner. To construct a regular satin or sateen weave (for irregular or special ones see Robinson and Marks or Watson) without a twill effect a number of rules have to be observed. The distribution of interlacing must be as random as possible and there has only to be one interlacing of each warp and weft thread per repeat, that is per weave number.

The intersections must be arranged in an orderly manner, uniformly separated from each other and never adjacent. The weaves are developed from a 1/x twill and the twill intersections are displaced by a fixed number of steps. The steps that must be avoided are: (i) one or one less than the repeat (because this is a twill), (ii) the same number as the repeat or having a common factor with the repeat (because some of the yarns would fail to interlace). The smallest weave number for either weave is 5 and regular satins or sateens also cannot be constructed with weave numbers of 6,9,11,13,14 or 15. The most popular weave numbers are 5 and 8 and weave numbers above 16 are impracticable because of the length of floats. Weave numbers of 2 or 3 are possible for five-end weaves and 3 or 5 for 8-end weaves.

TRIAXIAL WEAVES

Nearly all two-dimensional woven structures have been developed from plain weave fabrics and warp and weft yarns are interlaced at right angles or at nearly right angles. This also applies in principle to leno fabrics and to lappet fabrics where a proportion of the warp yarns, that is the yarns forming the design, are moved across a number of ground warp yarns by the lappet mechanism.

The only exceptions are triaxial fabrics, where two sets of warp yarns are generally inserted at 60° to the weft, and tetra-axial fabrics where four sets of yarns are inclined at 45° to each other. So far only weaving machines for triaxial fabrics are in commercial production. Triaxial weaving machines were first built by the Barber-Colman Co. under licence from Dow Weave and have been further developed by Howa Machinery Ltd., Japan.

Triaxial fabrics are defined as cloths where the three sets of threads form a multitude of equilateral triangles. Two sets of warp yarn are interlaced at 60° with each other and with the weft. In the basic triaxial fabric, the warp travels from selvedge to selvedge at an angle of 30° from the vertical. When a warp yarn reaches the selvedge it is turned through an angle of 120° and then travels to the opposite selvedge thus forming a firm selvedge. Weft yarn is inserted at right angles (90°) to the selvedge. The basic triaxial fabric is fairly open with a diamond-shaped centre. The standard weaves can be modified by having biplane, stuffed or basic basket weaves, the latter being. These modified weaves form closer structures with different characteristics. Interlacing angles of 75° to the selvedge can be produced. At present thread spacing in the basic weave fabric is limited to 3.6 or 7.4 threads per centimetre.

The tear resistance and bursting resistance of triaxial fabrics is greatly superior to that of standard fabrics because strain is always taken in two directions. Their shear resistance is also excellent because intersections are locked. They have a wide range of technical applications including sailcloths, tyre fabrics, balloon fabrics, pressure receptacles and laminated structures.

Selvedge

The selvedges form the longitudinal edges of a fabric and are generally formed during weaving. The weave used to construct the selvedge may be the same, or may differ from, the weave used in the body of the cloth. Most selvedges are fairly narrow but they can be up to 20-mm wide. Descriptions may be woven into the selvedge, using special selvedge jacquards, or coloured or fancy threads may be incorporated for identification purposes. For some end-uses selvedges have to be discarded but, whenever possible, selvedges should be constructed so that they can be incorporated into the final product so as to reduce the cost of waste.

In cloths woven on weaving machines with shuttles selvedges are formed by the weft turning at the edges after the insertion of each pick. The weft passes continuously across the width of the fabric from side to side. In cloths woven on shuttleless weaving machines the weft is cut at the end of each pick or after every second pick. To prevent the outside ends from fraying, various selvedge motions are used to bind the warp into the body of the cloth or edges may be sealed. The essential requirements are that the selvedges should protect the edge of the fabric during weaving and subsequent processing, that they should not detract from the appearance of the cloth and that they should not interfere with finishing or cause waviness, contraction or creasing. Four types of woven selvedges.

HAIRPIN SELVEDGE - SHUTTLE WEAVING MACHINE

A typical hairpin selvedge which is formed when the weft package is carried in a reciprocating media, for example a shuttle. With most weaves it gives a good edge and requires no special mechanism. Frequently strong two-ply yarns are incorporated into the selvedge whilst single yarns are used in the body of the cloth because the edge threads are subjected to special strains during beat-up. To ensure a flat edge a different weave may be used in the selvedge from the body of the cloth. For twills, satins and fancy weaves this may also be necessary to ensure that all warp threads are properly bound into the edge. If special selvedge yarns are used it is important to ensure that they are not, by mistake, woven into the body of the fabric because they are likely to show up after finishing or cause a reduction in the tear and/or bursting resistance of industrial fabrics.

When two or more different weft yarns are used in a fabric only one weft yarn is being inserted at a time and the other yarn(s) are inactive until the weft is changed. In shuttle looms the weft being inserted at any one time will form a normal selvedge whilst the weft yarn(s) not in use will float along the selvedge. If long floats are formed in weaving, because one weft is not required for a considerable period, the floats may have to be trimmed off after weaving to prevent them from causing problems during subsequent processing. Whenever a pirn is changed, or a broken pick is repaired, a short length of yarn will also protrude from the selvedge and this has to be trimmed off after weaving.

Industrial fabrics with coarse weft yarn are sometimes woven with loop selvedges to ensure that the thread spacing and cloth thickness remains the same right to the edge of the cloth. To produce a loop selvedge a wire or coarse monofilament yarn is placed 3 or 4 mm outside the edge warp end and the pick reverses round the wire to form a loop. As the wire is considerably stiffer than a yarn it will prevent the weft pulling the end threads together during beat-up.

The wire extends to the fell of the cloth and is automatically withdrawn during weaving. Great care is required to ensure that no broken-off ends of wire or monofilament remain in the fabric because they can cause serious damage to equipment and to the cloth during subsequent processing.

Tucked-in Selvedges

A tuck or tucked-in selvedge. It gives a very neat and strong selvedge and was first developed by Sulzer Brothers (now Sulzer-Textil), Win-terthur, Switzerland, for use on their projectile weaving machines. Its appearance is close to that of a hairpin selvedge and it is particularly useful when cloths with fringe selvedges would have to be hemmed. High-speed tucking motions are now available and it is possible to produce tucked-in selvedges even on the fastest weaving machines.

Tucked-in selvedges, even when produced with a reduced number of warp threads, are generally slightly thicker than the body of the cloth. When large batches of cloth with such selvedges are produced it may be necessary to traverse the cloth on the cloth roller to build a level roll. The extra thickness will also have to be allowed for when fabrics are coated. When tucked-in selvedges can be incorporated into the finished product no yarn waste is made because no dummy selvedges are produced but the reduced cost of waste may be counterbalanced by the relatively high cost of the tucking units.

Sealed Selvedges

When fabrics are produced from yarns with thermoplastic properties the edge of a fabric may be cut and sealed by heat. The edge ends of fringe selvedges are frequently cut off in the loom and the edge threads with the fringe are drawn away into a waste container. Heat cutting may also be used to slit a cloth, in or off the loom, into a number of narrower fabrics or tapes.

For special purposes ultrasonic sealing devices are available. These devices are fairly expensive and can cut more cloth than can be woven on one loom. Whilst they can be mounted on the loom they are, because of their cost, more frequently mounted on a separate cutting or inspection table.

Leno and Helical Selvedges

In most shuttleless weaving machines a length of weft has to be cut for every pick. For looms not fitted with tucking motions the warp threads at the edges of the fabric have to be locked into position to prevent fraying. With some weft insertion systems the weft has to be cut only after every second pick and it is possible to have a hairpin selvedge on one side of the fabric and a locked selvedge on the second side. Leno and helical selvedges are widely used to lock warp yarns into position respectively. When shuttleless weaving machines were first introduced there was considerable customer resistance to fringe selvedges but, in the meantime, it has been found that they meet most requirements.

With leno selvedges a set of threads at the edge of the fabric is interlaced with a gauze weave which locks round the weft thread and prevents ravelling of the warp. As the precut length of the picks always varies slightly, a dummy selvedge is sometimes used at the edge of the cloth. This makes it possible to cut the weft close to the body of the cloth resulting in a narrow fringe which has a better appearance than the selvedge from which weft threads of varying lengths protrude. It also has the advantage in finishing and coating that there are no long lengths of loose weft that can untwist and shed fly.

Because of the tails of weft and because of the warp in dummy selvedges, more waste is generally made in narrow shuttleless woven fabrics than in fabrics woven on shuttle looms. For wide cloth the reverse frequently applies because there is no shuttle waste. Leno selvedges, sometimes referred to as 'centre selvedges', may be introduced into the body of the cloth if it is intended to slit the width of cloth produced on the loom into two or more widths either on the loom or after finishing. Helical selvedges consist of a set of threads which make a half or complete revolution around one another between picks. They can be used instead of leno selvedges and tend to have a neater appearance.

REMOVING KNITTING LUBRICANTS FROM FABRIC

BAT is to do one of following:

- Select knitted fabric that has been processed using water-soluble and biodegradable lubricants instead of the conventional mineral oil-based lubricants. Remove them by water washing. With knitted fabrics made of synthetic fibres the washing step needs to be carried out before thermofixation (to remove the lubricants and avoid them being released in the form of air emissions)
- Carry out the thermofixation step before washing and treat the air emissions generated from the stenter frame by dry electrofiltration systems that allow energy recovery and separate collection of the oil. This will reduce the contamination of the effluent
- Remove the non-water soluble oils using organic solvent washing. The requirements are then taken, along with provisions for the in-loop destruction of the persistent pollutants (*e.g.* by advanced oxidation processes). This will avoid any possible contamination of groundwater arising from diffuse pollution and accidents. This technique is convenient when other non water-soluble preparation agents, such as silicone oils, are present on the fabric.

DESIZING

BAT is to do one of the following:

- Select raw material processed with low add-on techniques and more effective bioeliminable sizing agents combined with the use of efficient washing systems for desizing and low F/M waste water treatment techniques to improve the bioeliminability of the sizing agents
- Adopt the oxidative route when it is not possible to control the source of the raw material
- Combine desizing/scouring and bleaching in one single step.
- Recover and re-use the sizing agents by ultrafiltration.

BLEACHING

BAT is to:

- Use hydrogen peroxide bleaching as preferred bleaching agent combined with techniques for minimising the use of hydrogen peroxide stabilisers, or using biodegradable/bioeliminable complexing agents
- Use sodium chlorite for flax and bast fibres that cannot be bleached with hydrogen peroxide alone. A two-step hydrogen peroxide-chlorine dioxide bleaching is the preferred option. It must be ensured that elemental chlorine-free chlorine dioxide is used. Chlorine-free chlorine dioxide is produced using hydrogen peroxide as the reducing agent of sodium chlorate
- Limit the use of sodium hypochlorite only to cases in which high whiteness has to be achieved and to fabrics that are fragile and would suffer depolymerisation. In these special cases, to reduce the formation of hazardous AOX, sodium hypochlorite bleaching is carried out in a two-step process in which peroxide is used in the first step and hypochlorite in the second. Effluent from hypochlorite bleaching is kept separate from the other streams and mixed effluents in order to reduce formation of hazardous AOX.

DYEING

BAT is to do all the following:

- Reduce the number of dyes (one way to reduce the number of dyes is by using trichromatic systems)
- Use automated systems for dosage and dispensing of dyes, only considering manual operation for dyes that are used infrequently
- In long continuous lines where the dead volume of the distribution line is comparable with the volume in the padder, give preference to decentralised automated stations that do not premix the different chemicals with the dyes before the process and that are fully automatically cleaned.

GENERAL BAT FOR BATCH DYEING PROCESSES

BAT is to:

- Use machinery fitted with: automatic controllers of fill volume, temperature and other dyeing cycle parameters, indirect heating & cooling systems, hoods and doors to minimise vapour losses
- Choose the machinery that is most fitted to the size of the lot to be processed to allow its operation in the range of nominal liquor ratios for which it is designed. Modern machines can be operated at approximately constant liquor ratio whilst being loaded at a level as low as 60 % of their nominal capacity (or even 30 % of their nominal capacity with yarn dyeing machines)
- Select new machinery according as far as possible to the requirements:
  - Low- or ultra-low liquor ratio
  - In-process separation of the bath from the substrate
  - Internal separation of process liquor from the washing liquor
  - Mechanical liquor extraction to reduce carry-over and improve washing efficiency
  - Reduced duration of the cycle.
- Substitute overflow-flood rinsing method in favour of drain and fill or other methods (smart rinsing for fabric)
- Re-use rinse water for the next dyeing or reconstitution and re-use the dye bath when technical considerations allow. This technique is easier to implement in loose fibre dyeing where top-loading machines are used. The fibre carrier can be removed from the dyeing machine without draining the bath. However, modern batch dyeing machines are equipped with built-in holding tanks allowing for uninterrupted automatic separation of concentrates from rinsing water

MERCERISING

BAT is to either:

- Recover and re-use alkali from mercerising rinsing water
- Or re-use the alkali-containing effluent in other preparation treatments.

BAT FOR CONTINUOUS DYEING PROCESSES

Continuous and semi-continuous dyeing processes consume less water than batch dyeing, but highly concentrated residues are produced.

BAT is to reduce losses of concentrated liquor by:

- Using low add-on liquor application systems and minimising volume capacity of the dip trough when using pad dyeing techniques
- Adopting dispensing systems where the chemicals are dispensed on-line as separate streams, being mixed only immediately before being fed to the applicator

- Using one of the following systems for dosing the padding liquor, based on measurement of the pick up:
  - Measure the amount of dyeing liquor consumed by reference to the quantity of processed fabric (length of the fabric multiplied by its specific weight); the resulting values are automatically processed and used for the preparation of the next comparable batch
  - Use the rapid batch dyeing technique, where rather than being prepared for the whole batch before starting the dyeing batch, the dyestuff solution is prepared just in time, in several steps, based on on-line measurement of the pick up. This second technique is preferred when economic considerations allow
- Increase washing efficiency according to the principles of counter-current washing and reduction of carry-over.

PES & PES BLENDS DYEING WITH DISPERSE DYES

BAT is to:

- Avoid the use of hazardous carriers by (in order of priority):
  - Using non-carrier dyeable polyester fibres (modified PET or PTT-type), when product market considerations allow
  - Dyeing in HT conditions without use of carriers. This technique is not applicable to PES/WO and elastane/WO blends
  - Substituting conventional dye carriers with compounds based on benzylbenzoate and N-alkylphthalimide, when dyeing WO/PES fibres
- Substitute sodium dithionite in PES aftertreatment, by applying one of the two proposed techniques:
  - Replace sodium dithionite with reducing agent based on sulphinic acid derivatives. This should be combined with measures in order to ensure that only the strict amount of reducing agent needed to reduce the dyestuff is consumed (*e.g.* by using nitrogen to remove oxygen from the liquor and from the air in the machine)
  - Use of disperse dyes that can be cleared in alkaline medium by hydrolitic solubilisation instead of reduction
- Use optimised dye formulations that contain dispersing agents with high degree of bioeliminability.

BATCH DYEING WITH REACTIVE DYES

BAT is to:

- Use high-fixation, low-salt reactive dyes
- Avoid the use of detergents and complexing agents in the rinsing and neutralisation steps after dyeing, by applying hot rinsing integrated with recovery of the thermal energy from the rinsing effluent.

PAD-BATCH DYEING WITH REACTIVE DYES

BAT is to use dyeing techniques that perform at equivalent levels. The technique described is more cost effective than pad-batch dyeing in terms of total processing costs, but the initial capital investment in switching to the new technology is significant. However, for new installations and those seeking to replace equipment the cost factor is not so significant. In all cases, BAT is to avoid the use of urea and to use silicate-free fixation methods.

DYEING WITH SULPHUR DYES

BAT is to:

- Replace conventional powder and liquid sulphur dyes with stabilised non-pre-reduced sulphide-free dyestuffs or with pre-reduced liquid dye formulations with a sulphide content of less than 1 %
- Replace sodium sulphide with sulphur-free reducing agents or sodium dithionite, in that order of preference
- Adopt measures to ensure that only the strict amount of reducing agent needed to reduce the dyestuff is consumed (*e.g.* by using nitrogen to remove oxygen from the liquor and from the air in the machine)
- Use hydrogen peroxide as preferred oxidant.

WOOL DYEING

BAT is to:

- Substitute chrome dyes with reactive dyes or, where not possible, use ultra-low chroming methods that fulfil all the following requirements:
  - An emission factor of 50 mg chromium per kg of wool treated is achieved, which corresponds to a chromium concentration of 5 mg/l in the spent chroming bath when a 1:10 liquor ratio is used
  - No chromium (VI) is detectable in the waste water (using a standard method able to detect Cr VI at concentrations <0.1 mg/l)
- Ensure minimum discharge of heavy metals in the waste water when dyeing wool with metal complex dyes. BAT associated values are emission factors of 10 - 20 mg/kg of treated wool, which correspond to 1 - 2 mg/l of chromium in the spent dye bath when a 1:10 liquor ratio is used. These performances can be achieved by:
  - Using auxiliaries that enhance dye uptake like, for instance, the process for loose wool and tops
  - Using pH control methods to maximise final bath exhaustion for other make-ups
- Give preference to a pH-controlled process when dyeing with pH-controllable dyes (acid and basic dyes) so that level dyeing is obtained

with maximum exhaustion of dyes and insect resist agents and minimum use of organic levelling agents.

NONWOVEN BONDING PROCESSES IN BONDING TECHNOLOGY

Web consolidation or nonwoven bonding processes interlock preferentially arranged fiber or film assemblies by mechanical, chemical, solvent, and/or thermal means. The degree of bonding is a primary factor in determining fabric integrity (strength), porosity flexibility, softness, and density (loft, thickness). Bonding may be carried out as a separate and distinct operation, but generally is carried out as a sequential operation in tandem with web formation. In some fabric constructions, more than one bonding process may be used to enhance physical or chemical properties.

Mechanical consolidation methods include needlefelting, stitchbonding, and hydroentangling. Chemical consolidation methods involve applying adhesive binders to webs by saturating, spraying, printing, or foaming techniques. Solvent bonding involves softening or partially dissolving fibers with a solvent to provide self-bonding surfaces. Thermal bonding involves the use of heat and often pressure to fuse or weld fibers together at points of intersection or in patterned bond sites.

Important issues to consider when choosing the web consolidation methods are economy; versatility; and product properties, primarily absorbency, strength, softness, loft, and purity.

A recurring issue involves environmental requirements of both the process and the product. Many techniques are done for specific properties of unique fabrics; therefore, it is difficult to measure differences in cost. In some instances, two or more bonding techniques compete. The system that is most energy-efficient; environmentally sound or provides the preferred fabric properties generally dominates.

CHEMICAL BONDING PROCESSES

Chemical or resin bonding is a generic term for interlocking fibers by the application of a chemical binder. The chemical binder most frequently used to consolidate fiber webs today is a water-borne latex. Most latex binders are made from vinyl materials, such as polyvinylacetate, polyvinylchloride, styrene/butadiene resin, butadiene, and polyacrylic, or their combinations. Latexes are extensively used as nonwoven binders, because they are economical, versatile, easily applied, and effective adhesives. The versatility of a chemical binder system can be indicated by enumerating a few factors that are considered when such a system is formulated.

The chemical composition of the monomer or backbone material determines stiffness/softness properties, strength, water affinity (hydrophilic/hydrophobic balance), elasticity, durability, and aging. The type and nature of functional side groups determines solvent resistance, adhesive characteristics, and cross-linking nature. The type and quantity of surfactant used influences the polymerization process, polymer stability, and the application method.

Chemical binders are applied to webs in amounts ranging from about 5% to as much as 60% by weight. In some instances, when clays or other weighty additives are included, add-on levels can approach or even exceed the weight of the web. Waterborne binders are applied by spray, saturation, print, and foam methods. A general objective of each method is to apply the binder material in a manner sufficient to interlock the fibers and provide fabric properties required of the intended fabric usage. The common methods of bonding include saturation, foam, spray, print and powder bonding. They are briefly introduced in the following paragraphs:

Saturation

Saturation bonding is used in conjunction with processes which require rapid binder addition, such as card-bond systems, and for fabric applications which require strength, stiffness, and maximum fiber encapsulation, such as carrier fabrics. Fiber encapsulation is achieved by totally immersing the web in a binder bath or by flooding the web as it enters the nip point of a set of pressure rolls. Excess binder is removed by vacuum or roll pressure.

Three variations of saturation bonding exist: screen, dip/squeeze, and size-press. Screen saturation is used for medium-weight nonwovens, such as interlinings. Dip/squeeze saturation is used for web structures with strength sufficient to withstand immersion without support, such as spunbonds. Size-press saturation is used in high speed processes, such as wet-laid nonwovens. Drying and curing may be carried out on steam-heated drying cans or in thru-air ovens or perforated-drum dryers.

Binder addition levels range from 20% to 60%. Two techniques, single screen saturator and applicator roll technique. Advantages of this method are simplicity, controllable tensile strength and softness by choice and amount of binders. The disadvantages are the great influence of binders on softness, and the limitation in loftiness.

Foam Bonding

Foam bonding is a means to apply binder at low water and high binder-solids concentration levels. The basic concept employed involves using air as well as water as the binder diluent and carrier medium. Foam-bonded nonwovens require less energy in drying, since less water is used. The foam is generated by introducing air into the formulated latex while mechanically agitating the binder solution.

Air/latex dilutions or blow ratios in the order of 5:25 are practiced for various products. With the addition of a stabilizing agent to the binder solution, the foam can resist collapsing during application and curing, and the bonded fabric will exhibit enhanced loft, hand, and resilience. Non-stabilized foams are referred to as froths; froth-bonded fabrics are similar in properties to some saturation-bonded nonwovens. One example of this bonding. The advantages include less energy required to dry the web, less binder migration and controllable softness by choices and amount of binders. The disadvantages are difficulties in controlling process and adequate foaming.

Spray Bonding

In spray bonding, binders are sprayed onto moving webs. Spray bonding is used for fabric applications that which require the maintenance of highloft or bulk, such as fiberfill and air-laid pulp wipes. The binder is atomized by air pressure, hydraulic pressure, or centrifugal force and is applied to the upper surfaces of the web in fine droplet form through a system of nozzles.

Lower-web-surface binder addition is accomplished by reversing web direction on a second conveyor and passing the web under a second spray station. After each spraying, the web is passed through a heating zone to remove water, and the binder is cured (set/cross-linked) in a third heating zone. For uniform binder distribution, spray nozzles are carefully engineered.

Fig.: Industrial spray bonding process

Print Bonding

Print bonding applies binder only in predetermined areas. It is used for fabric applications that require a part of the area of the fabric to be binder-free, such as wipes and coverstocks. Many lightweight nonwovens are print bonded. Printing patterns are designed to enhance strength, fluid transport, softness, hand, absorbency, and drape. Print bonding is most often carded out with gravure rolls. Binder addition levels are dependent on engraved area and depth as well as binder-solids level. Increased pattern versatility can be achieved with the use of rotary screen rolls. Drying and curing are carried out on heated drums or steam-heated cans.

In print bonding, high viscose binders are applied to limited, patterned areas. A prewet/prebond step is required for enough strength of webs. There are two types of printers: rotary screen and rotogravure printers. Binders are applied through a hollow applicator roll in rotary screen printer, while in rotogravure printer they are applied by an engraved applicator roll. The main advantage is that outstanding softness of nonwoven fabrics with adequate strength can be achieved.

Powder Bonding

In powder bonding, the adhesive powder of thermoplastic polymers is applied onto webs by heat and pressure. Polyesters and polyolefins with low Tg's and molecular weight can be used as powder binders. A typical bonding line. The advantages are the bulky structure of dense nonwovens and the applicability of polyester or polypropylene webs. The disadvantage lies in difficulties of suitable particle sizes and ranges, and their distribution.

APPLICATIONS

Nonwoven products in which binders are utilized:

- Wipes and towels
- Medical nonwovens
- Roofing products
- Apparel interlinings
- Filter media
- Coating substrates
- Automotive trim
- Carrier fabrics
- Bedding products (high loft)
- Furniture applications (high loft)
- Apparel
- Pillows (high loft).

NONWOVENS TECHNOLOGY IN CHEMICAL BONDING

For more than four decades, almost all nonwovens required a chemical binder in order to provide any measure of structural integrity. In addition, the binder was called upon to contribute and convey numerous properties that were necessary for the effective performance of the fabric. During this extended period, binders were essentially the weak element in developing fully acceptable nonwoven fabrics.

The fibers that were available to the nonwoven industry were the same fibers that were available to the textile and other fiber-based industries; hence, the fibers were fully acceptable. Generally, the binder limited the performance of the nonwoven fabric. The deficiencies cited against nonwovens generally were deficiencies attributable to an inadequate binder. Common complaints are as follows:

The fabric doesn't have enough strength:

- The fabric is too stiff.
- The fabric has inadequate absorbency.
- The fabric shows poor laundering ability.
- The fabric has inadequate dry cleaning ability.
- The fabric simply doesn't feel like a textile.

Consequently, a great deal of effort has been put into the development and continuous improvement of chemical binders. The steady improvements in nonwovens performance that occurred over a period of many years were, in no small measure, due to improvements in the performance and utility of the binder. In the very early stages of nonwovens development, different types of natural resins and glues were used to bond nonwovens. While they conveyed some integrity and strength to these webs, they also had many glaring deficiencies. Consequently, synthetic binders were developed to meet the structural and performance requirements of nonwoven fabrics.

Polyvinyl acetate was the first successful synthetic binder used in substantial volume. This material had distinctly superior adhesive properties, strength, and performance compared to the early natural adhesives. This binder is flexible and it can be applied to fiber webs by many ways including print bonding. The industry was faced with the inevitable compromise in fabric properties of nonwovens bonded with synthetic materials. In order to build strength in the fabric, increasing amounts of resin must be applied, which results in more stiffness. If softness is necessary, it can be achieved, but primarily by sacrificing strength.

A substantial improvement in this trade-off of strength and softness was achieved with the introduction of acrylic-based latex binders in the 1950s and 1960s. By proper selection of co-monomers, it is possible to build improved softness properties with adequate strength. Consequently, these binders became widely used by most of the nonwovens industry, despite the somewhat higher cost. As polymer technology for manufacturers of synthetic binder systems improved, a greater variety of chemical building blocks became available with much greater flexibility in terms of binder strength, durability, and other properties. The introduction of cross-linkable and self-crosslinking binder polymers turned out an entirely new range of fabric properties. This was particularly noteworthy in durable nonwovens where such durability features as washability and dry cleanability were important.

HOW BINDERS WORK

The process involves three steps:

1. Binder application to nonwoven web.
2. Removal moisture or solvent.
3. Formation of strong bond between binder and nonwoven web.

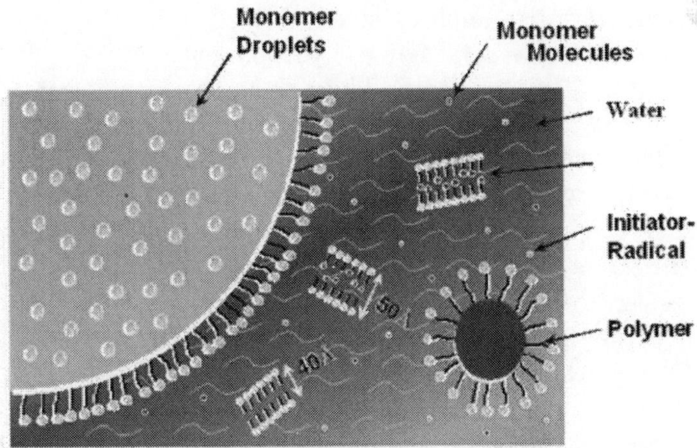

Fig. Emulsion Polymerization – Schematic

In general binders contain polymer produced by the reaction of monomers in presence of initiators or catalysts. Surfactants are used to stabilize these polymer particles in water during emulsification. After application of the binder to the nonwoven web, during moisture removal, film formation takes place.

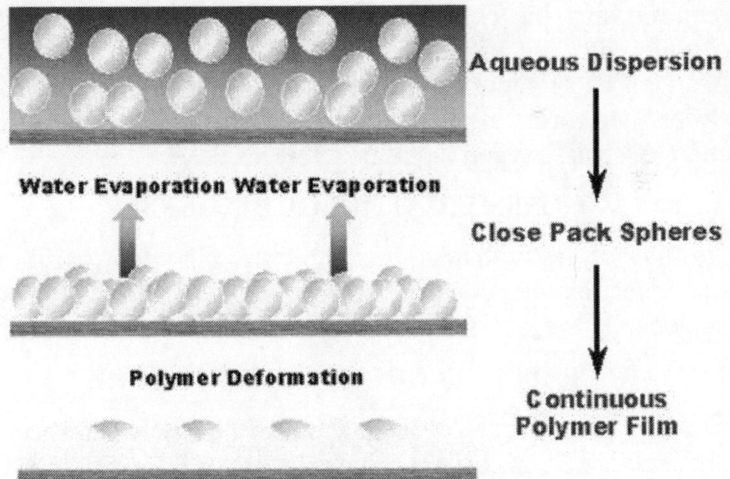

Fig. Schematic of Film Formation

Binder factors influencing nonwoven performance are:

1. Backbone Structure
2. Functional Group
3. Surfactant
4. Process

PROPERTIES DESIRED IN A BINDER

The construction of a nonwoven with suitable binders is to achieve improved characteristics such as strength, softness, adhesion, firmness, durability, stiffness, fire retardence, hydrophilicity, hydrophobicity, anti-microbial properties, organic compatibility, reduced surface tension, improved dimensional stability and solvent, wash and acid resistance. The following list illustrates some general considerations required for an ideal binder.

The required properties can be varied depending on the end-uses:

- Strength: The strength of a nonwoven fabric is more closely related to the strength of the applied binder.
- Adhesion to Fibers: Even though the mechanism of adhesion is not completely understood, the adhesion strength of the binder-to-fiber bond has to be considered.
- Flexibility/handle: The some movements of fibers should be allowed, especially when a soft hand is desired.
- Elastic Recovery: To avoid the permanent deformation of fabric, good elastic recovery is required under strain.
- Resistance to washing/ Drying cleaning: Some nonwoven products need durability in cleaning processes according to their end-uses.
- Good colour and colour retention: Diverse ranges of colours are required, and the colorfastness and yellowing problems should be considered.
- Economical: Minimizing the cost is an ongoing requirement.
- Other special requirements: Such as Flame resistance, resistance to chemicals, air, oxygen, light, heat, etc.

CLASSIFICATION OF BINDERS

Due to their diversity, binders may be classified into several categories based on polymer (binder) chemical structure, functionality and the type of curing reactions.

CLASSIFICATION BASED ON CHEMICAL STRUCTURE

There are three main kinds of binders: butadiene copolymers, acrylates, and vinyl copolymers. The chemical compositions influence T_g, hardness and softness, hydrophobicity and hydrophilicity, elasticity, aging, and dry tensile strength of binders. The higher the T_g, the higher will be the dry tensile strength of binders.

CLASSIFICATION BASED ON FUNCTIONALITY

The functionality of binders is in the functional groups attached to polymer chains, which influences wet and solvent properties. To modify binder properties, copolymerization with a small amount of monomers with special functionality is performed. The main functionalities in binders are carboxyl and amide side chains.

Carboxyl Functionality

This functionality is related to binders containing acrylic acid or methacrylic acid by copolymerization. The binders are crosslinkable since the functional group, carboxylic acid, provides sites for crosslinking reactions.

Amide Functionality

This functionality is related to binders containing acrylamide by copolymerization. The amide functionality provides crosslinking sites, and even the binders are self-crosslinkable.

N-metylol Amide (NMA) Functionality

This functionality is obtained after acrylamide is reacted with formaldehyde. The binders containing the substituted acrylamide groups have self-crosslinking properties and the possible reaction as follows:

CLASSIFICATION BASED ON TYPE OF CURING REACTIONS

The classification of reactivity refers to crosslinkability of binders, which is related to reaction with curing resins, crosslinking agents. The most common curing resin is melamine formaldehyde condensate resin involving reaction of n-methylol groups.

Non-crosslinkable Polymers

The polymers do not contain any of the functional groups. They cannot crosslink, even with external curing resins.

Crosslinkable Polymers

The polymers contain acid or amide functional groups. They can react with added curing resins, but the degree of crosslinking is limited.

Self-crosslinking Polymers

The polymers contain n-methylol functional groups. They can react with themselves, and a high crosslink density can be obtained by adding curing resins. Recent trends in chemical bonding: Although nonwoven manufacturers are seeking alternative technologies such as thermal bonding, chemical bonding still has its advantages and a promising market. Chemical bonding allows more room for fabric designs and fiber selections. Both disposable and durable products are supplied to roll goods producers and fiber manufacturers. On the environmental front, increasingly strict regulations and guidelines are driving a trend towards alternative products and technologies. Manufacturers and end-product suppliers alike are seeking ultra-low or formaldehyde-free binders. The growing consideration of the environmental impact of chemical binder and additives has become a focus of debate on the national and international level.

Latex Binder Chemical Types

A latex polymer consists of an aqueous medium with extremely fine liquid or solid polymer particles dispersed therein. The latex polymer generally is produced via free radical emulsion polymerization in water, whereby a vinyl monomer is combined with a small amount of other monomer (co-monomer) to create a high molecular weight polymer. The latex dispersion also will carry surfactants, stabilizers, and other additives to convey realistic properties to the latex itself. When used as a binder, the latex typically is combined with other components to provide the formulated binder ready for application to the fiber web. The formulated binder conveys many characteristics that are not possessed by the straight binder. Consequently, there is a substantial chemistry involved in combining the latex with the other components in order to prepare the formulated binder.

Binders are quite dependent upon the glass transition temperature (T_g) of the monomer unit selected to form the polymer. Differential Scanning Calorimeter (DSC) is used to determine T_g as shown below: The lower the (T_g) of the monomer units, the softer is the resulting polymer.

A sampling of the most common monomers used in the manufacture of latex polymer for nonwoven binders include the following materials:

- Monomer Tg(°C)
- Ethylene –125
- Butadiene –78
- Butyl Acrylate –52
- Ethyl Acrylate –22
- Vinyl Acetate +30
- Vinyl Chloride +80
- Methyl Methacrylate +105
- Styrene +105
- Acrylonitrile +130

The monomers selected for forming the polymeric latex also have considerable influence on the hydrophilic or hydrophobic nature of the binder. This can affect the wet strength of the nonwoven fabric as well as a host of absorbency characteristics. With the current capabilities of polymerization chemistry, there is considerable versatility for each chemical type. Despite this range of properties, the commonly employed nonwovens binders generally are characterized by a fairly well defined set of properties. These properties can be modified to some degree by incorporation of other agents, but they provide a useful guide in classifying the kind of performance to be expected from each type of binder.

TYPES OF BINDERS

The following comparison of latex binder chemical types provides an indication of the relative performance, as well as the advantages and disadvantages of each type of binder. As indicated, the binder properties can be modified considerably by the presence of co-monomers. Acrylic: These binders offer the greatest durability, colour stability, and dry/wet performance. Acrylic binders have the widest range of fabric hand properties. They can be formulated to vary from very soft (T_g = - 40°C) to extremely hard (Tg = 105°C). These binders can be used in virtually all nonwovens applications, although they tend to be more costly. These polymers can be made to cross-link, with substantial improvement in durability.

Styrenated Acrylics: These are tough, hydrophobic binders. The resulting textile hand ranges from soft-to-firm (T_g varies from –20°C to +105°C).These binders can be used in applications where there is a need for some wet strength without crosslinking. The use of this type of latex binder does involve some sacrifice in UV and solvent resistance. Vinyl Acetate (VAC): The vinyl acetate binders are firm (T_g = +30°C to +40°C); however, they are relatively low cost and find extensive use. They offer good dry strength and toughness, but are somewhat hydrophilic and have a tendency to yellow when subjected to heat.

Vinyl Acrylics: These binders are more hydrophobic than the straight VAC binders. They provide excellent toughness, flexibility, and better colour stability. They are the compromise between VAC and acrylic, and can compete on a cost/performance basis. The hand range is limited to intermediate softness (T_g = -10°C) to a firm hand (T_g = +30°C).

Ethylene Vinyl Acetate (EVA): These latex binders have a (T_g range of –20°C to +115°C, which is equivalent to soft ranging to an intermediate textile hand. They exhibit high wet strength, coupled with excellent absorbency. In general, they are less costly than acrylics. They do have a tendency to have more of an odor compared to other binders. They are used primarily in wipes, air-laid pulp fabrics and similar applications.

Styrene-Butadiene (S/B, SB, or styrene butadiene rubber): These binders have an excellent combination of flexibility and toughness. They range in hardness from very soft (T_g = -30°C) to very firm (T_g = +80°C). However, the (T_g of an SB binder is not strictly comparable to other classes of nonwoven binders. The styrene-to-butadiene ratio (S/B ratio) is the most common method for describing the relative hand resulting from the use of these binders. When cross-linked, this class of binder is very hydrophobic and durable. They are affected somewhat by heat and light because of their tendency to oxidize.

Polyvinyl Chloride (PVC): The homopolymer of polyvinyl chloride is a very hard, rigid polymer (T_g = +80°C). This polymer must be plasticized to provide flexibility and film-forming properties. Normally, the (PVC) binders used in nonwovens are softened internally by co-polymerizing the vinyl chloride or with softer acrylic monomers. The hand range of most of these polymers is still relatively firm (T_g is greater than the +30°C). Because this type of polymer is a thermoplastic, it performs well in heat and dielectric sealing applications. This can be an advantage in some uses.

The chlorine content of the polymer promotes flame retardency. This feature is one of the primary benefits of utilizing this type of binder. However, the chlorine also conveys the tendency to yellow upon heat aging, due to elimination of hydrogen chloride from the polymer.

Ethylene/Vinyl Chloride: Binders in this class have a slightly broader hand range (T_g = 0°C to +30°C) without the external plasticization required of (PVC) binders. The presence of the chlorine again conveys some flame retardancy. These binders exhibit good acid resistance, fair water resistance, and excellent adhesion to synthetic fibers. There is some tendency to yellow upon aging. In essence, this is an internally plasticized (PVC) binder, considering the ethylene monomer to be the softener.

FORMULATION

Ingredients

The formulation of binding solution is an art since many ingredients are involved and many different possibilities exist for different end-uses. Some of the characteristics, and the types of formulation agents utilized to obtain them include the following.

- Surfactants: offer improvement in binder adhesion, stability, and ability to be converted into a foam
- External cross-linkers: provide cross-links with binder polymer to provide improved performance
- Defoamers: utilized to minimize foam in processing
- Repellent agents: convey water or oil repellency
- Salts: added to impart low flame response properties and to convey antistatic properties
- Thickeners: added to control the rheology of the binder liquid
- Catalysts: added to facilitate curing and to promote cross-linking
- Acids and bases: added to control pH of the latex
- Dyes and pigments: provide colour to the binder and fabric
- Fillers: added to reduce binder tack and to lower cost
- Optical brighteners: added to increase whiteness
- Sewing aids: added to provide lubrication during fabrication

The purposes of wetting agents, mainly nonionic or anionic surfactants, are to enhance binder penetration through webs, improve the affinity between binder and fibers.

The crosslinker, which has multi-functional groups, is generally added to increase crosslink density and to improve durability and resistance to deformation.

Order of Formulation

In terms of adding ingredients into a binding bath, the compatibility of ingredients should be confirmed because the orders are extremely important. The milky white colour of most binders impedes a check on the white-colour indication of non-compatible ingredients, so most ingredients are first added to the dilution water. After the compatibility is assured, binders are added and then thickeners added to adjust viscosity. For the stability of the binding solution, catalysts are added just before application. Some water may be added to reach a desirable solid level.

The summarized order is as follows:

- Most ingredients
- Latex binder
- Thickener
- Catalyst
- Some water, and the others, such as dyes and pigments, fillers, clays, optical brighteners, sewing aids, etc.

KNITTING AND FASHION INDUSTRY

Knitted fabric has certain special characteristics that make it suitable for creating a wide range of garments and accessories like tights, gloves, underwear and other close-fitting garments. The structure of the loop of knitted fabric stretches and molds to fit body shapes. The air trapped by the interlocking loops keeps the wearer warm.

There are two important features of knitted fabric which can be related to fashion:

- Fashioning mark
- Fully-fashioned.

Increasing or decreasing the number of stitches in a row widens or narrows the garment being created while knitting. Increasing is attained by moving outer loops sideways on a frame and thereby creating extra loops. This process leaves a small eyelet hole in the fabric which is known as a fashioning mark. The process is reversed while decreasing and the loops move inwards. Here the fashioning mark appears where two loops are compressed into one new loop. Garments made in this way are known as fully-fashioned and are regarded as high quality. It is seen today that some firms used fake fashioning marks to make garments appear to be fully-fashioned.

The popularity of knitting has grown a lot within the recent years owing to the adaptability of various man-made fibers, the increased versatility of knitting techniques and the growth in demand for wrinkle-resistant, stretchable, snug-fitting fabrics (particularly in the range of sportswear and other casual apparels). Today knitted fabrics form an integral part of hosiery, underwear, slacks, sweaters, suits and coats, rugs and other home furnishing items.

TWO DIVISIONS OF THE KNITTING INDUSTRY

Knitting industry has two main divisions:

- One division produces knitted goods for apparel manufacturers, for sewing centres, for consumers and for others.
- Other division produces completed apparel like hosiery, sweaters and underwear.

Knitting is a method by which thread or yarn may be turned into cloth. Knitting consists of loops called stitches pulled through each other. The active stitches are held on a needle until another loop can be passed through them. Knitting may be done by hand or by machine. By hand, there are numerous styles and methods. Flat knitting, which is done on two straight needles, produces a length of cloth, while circular knitting, which is done on circular or double-pointed needles, produces a seamless tube. Different yarns and knitting needles may be used to achieve different end products, by giving the final piece different colour, texture, weight, or integrity.

KNITTING VS CROCHETING

Knitting is an ancient art of stitching and has been a popular technique of making sweaters, woollens, and stockings for many centuries. Though today, knitting is done by both machines and hand, hand knitting is still a popular hobby for many. Knitting requires a continuous strand of yarn and two or more needles. The two basic stitches of knitting - knit and purl, can be worked alone or together and form the basis of varieties of designs as well as other stitches. You begin knitting by making a slip knot on one needle and "casting on" the number of stitches you need for the project.

In this way you create the foundation row on one needle. The basic knitted stitches are created using both needles, wrapping the yarn over one needle and drawing the wrapped yarn through loops on the other needle. Besides the single point needle and double pointed needle, circular knitting needles are growing in popularity because they eliminate seams and the need to continually turn your knitting at the end of a row. Like Knitting, Crocheting is also a very popular technique for making a variety of fashion and home decor accessories. By combining lighter weight or softer yarns and basic crochet stitches, you create a delicate, drapable fabric. On the other hand a thicker yarn produces a sturdy fabric.Beautiful textured and raised stitches are easy to make in crochet. To do crocheting you require a continuous strand of yarn and a single hook. You start with a slip stitch and continue to make loops. These are called chains and using the chains you create a foundation row. Rows are built on this foundation. Crochet stitches are made with loops and yarn is continuously wrapped around the hook. The loops are drawn through the wrapped yarn to make the stitches. Crocheting can be done in rows, keeping your work flat, or you can join your stitches, creating a ring and work in the round.

A Comparison of Knitting and Crocheting

Knitting and crocheting are popular hobbies. These were once important home crafts necessary to keeping one's family clad and warm. Today, most of the knitted and crocheted garments are machine-made. But handcrafted items are also available at specialty shops or made by many at home both for fun and for the pride that comes from wearing something one made on his own. Knitting and crocheting are similar processes in the sense that loops of yarn or thread are pulled through existing loops to make chains and fabrics. However, the two crafts are very different in how they are accomplished.

Knit and Crochet Tools

Knitting uses two or more pointed needles. Crocheting requires only one needle with a hook on the end.

Knit and Crochet Stitches

Knitting and crocheting stitches are different in the sense that in crocheting, the crafter works on one loop at a time and makes chains by pulling a loop of yarn through the loop on the hook or inserts the hook into the existing fabric and pulling a loop through both the fabric and the loop on the hook, leaving only one loop on the hook. On the other hand, in knitting, multiple loops or stitches are worked across the needle, the width of the fabric being determined by the number of stitches across. However, there can be similar effects achieved in both knit and crochet. But they are just achieved differently. Cable stitch, the braided pattern that is a standard feature of a knit sweater, can even been done in crochet. Working in the round or crafting a tube of fabric is possible in both knitting and crocheting.

It is not necessary that a person who knows crocheting also knows knitting. But both the crafts are enjoyable, relaxing and meditative. Knitting or crocheting while watching your favourite television programs is a favourite pastime for many and which allows you to claim that you are really "working" and not just "wasting your time".

STRUCTURE

Courses and Wales

Structure of stockinette, a common knitted fabric. The meandering red path defines one *course*, the path of the yarn through the fabric. The uppermost white loops are unsecured and "active", but they secure the red loops suspended from them. In turn, the red loops secure the white loops just below them, which in turn secure the loops below them, and so on. Alternating wales of red and white knit stitches. Each stitch in a wale is suspended from the one above it. Like weaving, knitting is a technique for producing a two-dimensional fabric from a one-dimensional yarn or thread. In weaving, the threads are always straight, running parallel either lengthwise (warp threads) or crosswise (weft threads).

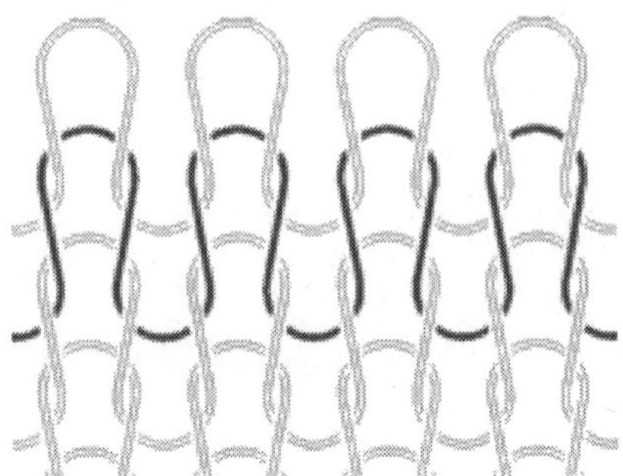

By contrast, the yarn in knitted fabrics follows a meandering path (a *course*), forming symmetric loops (also called bights or *stitches*) symmetrically above and below the mean path of the yarn. These meandering loops can be stretched easily in different directions, which gives knitting much more elasticity than woven fabrics; depending on the yarn and knitting pattern, knitted garments can stretch as much as 500%.

For this reason, knitting was initially developed for garments that must be elastic or stretch in response to the wearer's motions, such as socks and hosiery. For comparison, woven garments stretch mainly along one direction (the *bias*) and not very much, unless they are woven from stretchable material such as spandex. Knitted garments are often more form-fitting than woven garments, since their elasticity allows them to follow the body's curvature closely; by contrast, curvature is introduced into most woven garments only with sewn darts, flares, gussets and gores, the seams of which lower the elasticity of the woven fabric still further.

Extra curvature can be introduced into knitted garments without seams, as in the heel of a sock; the effect of darts, flares, etc. can be obtained with short rows or by increasing/decreasing the number of stitches. The thread used in weaving is usually much finer than the yarn used in knitting, which can make the knitted fabric bulkier and have less drape than a woven fabric. If they are not secured, the loops of a knitted course will come undone when their yarn is pulled; this is known as *ripping out* or *unravelling* knitting. To secure a stitch, at least one new loop is passed through it. Although the new stitch is itself unsecured ("active"), it secures the stitch(es) suspended from it. A sequence of stitches in which each stitch suspended from the next is called a *wale*. To secure the initial stitches of a knitted fabric, a method for casting on is used; to secure the final stitches in a wale, one uses a method of binding off. During knitting, the active stitches are secured mechanically, either from individual hooks (in knitting machines) or from a knitting needle or frame in hand-knitting.

Parallel white, red and green yarns zigzag *lengthwise* along the fabric, each loop securing a loop of an adjacent strand from the previous row. Thus, the two central wales in this picture are alternating *white-red-white* and *red-green-red* stitches.

Weft and Warp Knitting

The method of converting yarn into fabric by intermeshing loops, which are formed with the help of needles is known as knitting. There are two basic forms of knitting technology:

Weft Knitting and Warp Knitting

Weft knitting is done with hands, and with machines. Warp knitting is done by machines only. In weft knitting, only a single yarn is required and the fabric consists of horizontal parallel courses of yarn. On the other hand, warp knitting requires one yarn for every stitch in the row, that is the course, and these yarns make vertical parallel wales. Warp knitting is resistant to runs, and is commonly used as a lingerie fabric.

Weft knitted fabrics can be produced in either tubular or flat form. Weft knitted fabric is usually highly elastic and highly drapable, which makes it suitable for a wide range of apparel applications. The main advantage of warp knitted fabric is that it is not easy to unravel like weft knitted fabric. However, this fabric is not as elastic as weft knitted fabric. Let us now study in detail the two basic types of knitting:

WEFT KNITTING

The process of hand knitting is known as weft knitting and machines can also do it. In weft knitting, the work progresses back and forth, that is widthwise. In each cycle, which is known as course, a new row of stitches is formed. In each row there may be a number of stitches depending upon the width of the fabric to be knitted.

Each stitch of the row is built-up intermeshing with the previously held stitches of the previous row. The vertical row of stitches or loops hanging vertically from the needles is known as wale. Weft knitted fabric is produced in either flat or tubular form.

Types of Weft Knitted Stitches

There are three fundamental stitches in weft knitting:

- Plain-knit Stitch
- Purl Stitch
- Rib Stitch.

Plain-Knit Stitch

The basic form of knitting is the Plain Knit. This stitch can be produced in flat knit, tubular, or circular forms. The flat knit is also called jersey stitch. In plain knitted stitch, each loop is drawn through other loops to the right side of the fabric. The loops form vertical rows, also called wales, on the fabric face, giving it a sheen, and crosswise rows, also called courses, on the back. The plain-knit stitch produces relatively lightweight fabric compared with the thicker fabrics produced by other stitches.

Purl Stitch

This stitch is also known as the links/links stitch. It is made on flatbed and circular machines by needles using hooks on both ends. The hooks alternately draw loops to the front of the fabric in one course, and to the back in the next course. It is a slow and costly technique. Purl-stitched fabric looks the same on both sides and resembles the reverse of the plain knit. Because the purl stitch has crosswise stretch and excellent lengthwise stretch, it is widely used in kids' and infant wear.

Rib Stitch

The Rib knitted stitch is produced either on a flat rib machine or a circular rib machine. Rib knitted fabric has alternating rows of plain and purl stitches constructed so that both the face and back of the fabric look alike. Rib construction is costlier because of the greater amount of yarn needed. Weft Knitted Fabric is usually highly elastic and highly drapeable and these two properties make the fabric suitable for a wide range of apparel applications. Weft knitted fabric is considered to be comfortable both for outer garments and undergarments.

In the more common *weft knitting*, the wales are perpendicular to the course of the yarn; however, in warp knitting, the wales and courses run roughly parallel. In weft knitting, the entire fabric may be produced from a single yarn, by adding stitches to each wale in turn, moving across the fabric as in a raster scan. Weft-knit fabrics may also be knit with multiple yarns, usually to produce interesting colour patterns.

The two most common approaches are intarsia and double knitting. In intarsia, the yarns are used in well-segregated regions, *e.g.*, a red apple on a field of green; in that case, the yarns are kept on separate spools and only one is knitted at any time. In the more complex double knitting, two or more yarns alternate repeatedly within one row and all the yarns must be carried along the row, as seen in Fair Isle sweaters. Double knitting can produce two separate knitted fabrics simultaneously, *e.g.*, two socks; however, the two fabrics are usually integrated into one, giving it great warmth and excellent drape. In the knit stitch on the left, the next (red) loop passes through the previous (white) loop from *below*, whereas in the purl stitch (right), the next stitch enters from above. Thus, a knit stitch on one side of the fabric appears as a purl stitch on the other, and vice versa.

Warp Knitting

While a weft-knitted fabric consists of horizontal parallel courses of yarn and requires only a single yarn, warp knitting requires one yarn for every stitch in the row or course. The basic feature of warp knitting is that these yarns make vertical parallel wales. It differs from weft knitting in that each needle loops its own thread. The term 'warp knitting' refers to machine knitting. The needles produce parallel rows of loops simultaneously that are interlocked in a zigzag pattern. The stitches on the face of the fabric appear vertically and the stitches at the back appear horizontally as floats at a slight angle. These floats are also called Laps or Underlaps, and are a distinguishing identification of warp knits.

EVOLUTION OF KNITTING MACHINES

The first knitting machine invented in the 16th century by William Lee was meant to produce 8 loops per inch. Lee himself modified his machine within 9 years to produce 20 loops per inch. After this invention, the art of knitting was gradually taken over by guild-organized cottage industries. It is interesting to note that even today, the basic technology of the modern knitting machines is similar to Lee's machine. The full-fashioned machine was invented in 1864 by William Cotton of Leicestershire, England. He used the same bearded-spring needle which was a part of the original model of Lee. Automatic or power-operated knitting machines came into existence in the 19th century. Gradually, this upgraded to power-operated V-Bed, and circular knitting machines. Subsequently microprocessor-controlled knitting machines also came into existence.

Computerized Knitting

Computer-Aided Design (CAD) and Computer-Aided Manufacturing (CAM) have revolutionized the knitting industry. The 1970s saw the introduction of CAD/CAM systems. Designers used the CAD system to create product designs and these were transferred to CAM machines to manufacture the final product. CAD/CAM technology replaced the mechanical patterning and shaping devices on machines with electronic controls.

These systems enabled companies to respond quickly to changes in demand. New designs could be set up using CAD and quickly produced on the CAM machine. Though manually operated flat and V-bed knitting machines are still used, computerized technology is also preferred. Initially, this system was expensive and only the major companies could afford it. However, in the 1980s and 1990s CAD/CAM fell in price and even the small and medium sized companies invested in this new technology. The introduction of this technology enabled companies to work globally, and on a fast scale.

Introduction of Whole garment Knitting

Side by side with the CAD/CAM knitting technology, the process of whole garment knitting was introduced. Cutting fabric into shapes and then stitching the pieces together to create garments has been in practice since the eighteenth century. In the 1970s companies researched more efficient technologies which could produce a complete garment in one process without the need for sewing. Without the loss of fabric associated with cut and sew techniques, garments could be knitted quickly.

This was the "Whole garment Technology" which was further developed in the 1990s by Shima Seiki of Japan. Shima Seiki achieved this advancement in knitting technology by developing the slide needle. The slide needle was the result of research of the latch needle and the knitting process. Wholegarment knitted clothes provide greater comfort than cut and sew clothes. The absence of seams means that the garment can fit close to the body and stretch more easily. Wholegarment machines are also capable of producing complex designs. 2D knitted fabric is replaced by 3D knitwear, shaped and pleated as required.

Digital Printing on Knitted Fabric

With the advance of science and technology, textile printing on knitted garments has been dominated by screen printing technology. Advances in digital inkjet printing technology during the 1990's and early 2000's have resulted in knitting machines that can cost-effectively transfer designs directly from a computer and print onto fabric.

Latest Developments in Knitting

The art of knitting has come of age. From hand knitting to hand operated machines was a long way. Gradually it upgraded to power operated V-Bed and circular knitting, and then it leapfrogged to microprocessor controlled machines. Now lately CAD/CAM has revolutionised the knitting industry. However, in the developing countries including China, they still use manually operated flat and V-bed knitting machines for their bulk production since cheap labour is available. The other reasons being flexibility in knitting and economical for smaller design lots. However, quality on these machines remains a suspect because settings of stitch length and feed tensions may vary from machine to machine, which pose problems in sizing.

Styles of knitted fabrics

- Boiled wool
- Bunting
- Double knits
- Four-way stretch knits
- Interlock knits

- Jersey
- Milanese
- Power net
- Raschel knits
- Rib knits
- Single knits
- Stable knits
- Stretch knits
- Stretch velour
- Sweater knits
- Sweatshirt knits
- Tricot
- Two-way stretch knits

Methodology and Research Scope

The comparative analysis of the structure of knitted fabrics manufactured by the new and the traditional kinds of stitch The test material selected in the manner mentioned above included knitted fabric variants differentiated by stitch kind and warp take-up. In accordance with the designation codification accepted for the individual stitch variants, the designation Wn3tv (as an example) is related to the variant of a new stitch with sinker loop lengths which are characteristic of tricot (1st guide bar, upper warp), and velvet (2nd guide bar, bottom warp), manufactured at minimum warp take-up, and knitted in by both the first and second guide bars.

During the technological realisation of the designed knitted fabric variants, the thread tensions of the knitted-in warps were monitored to guarantee an appropriate value, not too high but not significantly lower than the average value. A special device for measuring the knitted-in warps was used. The device was equipped with a drum which, by rolling on the surface of the warp beam, measured the length of the thread segment knitted in 100 loop courses of the knitted fabric. While using the warp-knitting machine selected for manufacturing the test samples, we did not succeed in obtaining knitted fabrics of identical yarn lengths in the loop for the stitches of the new and traditional groups, as equivalents for comparison. The technological parameters for every stitch can be changed within certain limited ranges, whereas the groups of the stitches with mutually different location of the loops in the knitted fabrics' structure demand the use of different technologies. Nevertheless, the comparison of the new stitch variants with the equivalent traditional stitches allowed us to point out differences in their structure.

The comparative analysis of the area density values Pa showed that the new kind of stitches gives a significant higher loop number per surface square unit for all compared knitted fabric variants differentiated by stitch and warp take-up. The character of this parameter's changes in dependence on the systems of component stitches used is the same for both kinds of stitches.

The greatest area density Pa have stitches with the shortest sinker loops (group II), whereas the lowest have stitches with the longest sinker loops (group I). An alternative arrangement of component stitches results in intermediate values in relation to the above-mentioned for both kinds of stitches, but these are greater for stitch group IV (as for group III) with the longer sinker loop in the stitch being manufactured by the 1st guide bar.

The greatest value differentiation of the parameter discussed between new and traditional stitches occurs for stitch group II, and the smallest for group III. bottom warp (2 and 2'), as well as for both kinds of stitches. Between the values of the cover factors Zl related to the knitted fabric variants with maximum (2 and 2') and minimum (3 and 3') warp take-up, the values of cover factors for knitted fabrics with differentiated warp take-up values (4, 5, 4', and 5') are located.

The values of the area cover factor Za obtained for the knitted fabrics analysed show that the greater area filling of the knitted fabrics with yarn is characterised by stitches of the new group for all stitch variants. The highest Za values occur when the same component stitches (of the longest sinker loops) are applied for both kinds of stitches (group I). The analysis of the Za values for variants of groups III and IV proved that a higher area filling of the knitted fabrics with yarn can be obtained by applying a longer sinker loop of the component stitch in the 1st guide bar (group IV) for both kinds of knitted fabrics. The increase in the warp take-up lengths (3, 1, 2, 3', 1', and 2') is always associated by a lowering of the Za factor values for both stitch kinds.

Analysing the knitted fabric variants of all stitch groups for the maximum Za values, we can state that stitches of the new group have an area factor greater by 25% in relation to the traditional stitches. The higher values of the linear (Zl) and the area (Za) cover factors for the new family of warp-knitted stitches undoubtedly introduce significant changes in the use properties of knitted fabrics manufactured by means of this new kind of stitches. The greater filling here is not only the consequence of applying relatively smaller yarn segments taken up in the loops, but is above all the result of a different placement of loops in the stitch structure.

The values of the volume cover factor for stitches of the new group are higher than for the traditional stitches. The increase in the sinker loop length is associated with a lowering of the Zv values in the case of new stitches, whereas for traditional stitches a small increase occurs bottom warp (2 and 2'), as well as for both kinds of stitches. Between the values of the cover factors Zl related to the knitted fabric variants with maximum (2 and 2') and minimum (3 and 3') warp take-up, the values of cover factors for knitted fabrics with differentiated warp take-up values (4, 5, 4', and 5') are located.

The values of the area cover factor Za obtained for the knitted fabrics analysed show that the greater area filling of the knitted fabrics with yarn is characterized by stitches of the new group for all stitch variants. The highest Za values occur when the same component stitches (of the longest sinker loops) are applied for both kinds of stitches (group I). The analysis of the Za values for variants of groups III and IV proved that a higher area filling of the knitted fabrics with yarn can be obtained by applying a longer sinker loop of the component stitch in the 1st guide bar (group IV) for both kinds of knitted fabrics. The increase in the warp take-up lengths (3, 1, 2, 3', 1', and 2') is always associated by a lowering of the Za factor values for both stitch kinds.

Analysing the knitted fabric variants of all stitch groups for the maximum Za values, we can state that stitches of the new group have an area factor greater by 25% in relation to the traditional stitches. The higher values of the linear (Zl) and the area (Za) cover factors for the new family of warp-knitted stitches undoubtedly introduce significant changes in the use properties of knitted fabrics manufactured by means of this new kind of stitches. The greater filling here is not only the consequence of applying relatively smaller yarn segments taken up in the loops, but is above all the result of a different placement of loops in the stitch structure. The values of the volume cover factor for stitches of the new group are higher than for the traditional stitches. The increase in the sinker loop length is associated with a lowering of the Zv values in the case of new stitches, whereas for traditional stitches a small increase occurs.

7

Silk Industry

Silk, fibre produced as a cocoon covering by the silkworm, and valuable for its use in fine fabrics and textiles. The silkworm, in fact, is not a worm but a caterpillar. Although cocoon coverings of fibre are made by a large number of insects, only those of the mulberry silk moth, *Bombyx mori,* and a few other moths closely akin to it, are used by the silk industry. The silk of other insects, notably the spider, is used for certain manufacturing purposes, particularly for the cross hairs of telescopes and other optical instruments. For a description of the life and habits of the silk moth.

ORIGIN OF SILK

Silk is one of the oldest known textile fibres and, according to Chinese tradition, was used as long ago as the 27th century BC. The silkworm moth was originally a native of China, and for about 30 centuries the gathering and weaving of silk was a secret process, known only to the Chinese. Tradition credits Hsi-ling-shi, the 14-year-old bride of the Emperor Huang Ti, with the discovery of the potential of the cocoon and the invention of the first silk reel. China successfully guarded the secret until AD300, when Japan, and later India, penetrated the secrecy.

References in the Old Testament indicate that silk was known in biblical times in western Asia, from which it was presumably transplanted to the Greek Islands of the Aegean Sea. The Chinese are believed to have built up a lucrative trade with the West from the days of the Han dynasty in the 2nd century BC. The ancient Persian courts used Chinese silks, unraveled and rewoven into Persian designs.

When Darius III, king of Persia, surrendered to Alexander the Great, he was clothed in such silken splendor that Alexander was completely overshadowed and demanded as spoils the equivalent of $7 million in silk.

Caravans carried silk on camelback from the heart of Asia to Damascus, Syria, the marketplace at which East and West met. Here silk was traded for Western luxuries, some of which survive in China today. Silk became a valuable commodity in both Greece and Rome.

The Roman statesman and general Gaius Julius Caesar restricted silk to his exclusive use and to use for the purple Roman stripes on the togas of officials he favoured. Despite this, however, the use of silk in Rome spread in the era of pomp and display.

Until AD550 all silk woven in Europe was derived from Asiatic sources. About that time, however, the Roman emperor Justinian I sent two Nestorian monks to China, where, at the risk of their lives, they stole mulberry seeds and silkworm eggs, secreted them in their walking staffs, and brought them to Byzantium. Thus, the Chinese and Persian silk monopolies ended. With the spread of Islam, the silkworm came to Sicily and Spain.

By the 12th and 13th centuries Italy had become the silk centre of the West, but by the 17th century France was challenging Italy's leadership, and the silk looms established in the Lyons area at that time are still famous today for the unique beauty of their weaving. Not until after the revocation of the Edict of Nantes did Huguenot weavers from France cross the English Channel and establish silk mills in Spitalfields, a district in the East End of London. The silkworm, however, did not flourish in the English climate, nor has it ever flourished in the United States. The first silk mill in the U.S. was erected in 1810. With the advent of the power loom and with the help of the high tariffs, introduced during the American Civil War, against imported woven goods, the American silk-weaving industry entered a period of growth. Only in China, Japan, and, to a lesser extent, Italy and France was the silk itself produced, however.

In addition to the true silk from the cultivated mulberry silkworm, a number of so-called wild silks are produced from other related species of insects in the uncultivated state. Tussah silk is produced from a species that feeds on oak leaves. Douppioni is a silk produced from two silkworms that spin a cocoon together and thus produce a double thread. Special types of weaves, such as shantung and other irregular types, are woven from these types of silk.

PRODUCTION

The advent of synthetic fibres such as nylon and polyester, which are stronger than silk and lower in price, but do not possess the same hand, or quality, has caused a tremendous reduction in silk production and consumption. World production in 1940 was 59 million kg (130 million lb). By 1950 it had dropped to 19 million kg (42 million lb), but by the mid-1980s had climbed to about 68 million kg (150 million lb).

Silk is still used for clothing, including lightweight suits, coats and slacks, jackets, shirts and neckties, robes, loungewear, underwear, hosiery, and gloves. Silk is also used in lace, napery, draperies, linings, narrow fabrics, and handbags.

SERICULTURE

Sericulture, or the raising of silkworms, involves the incubation of the tiny eggs of the silkworm moth until they hatch and become worms. After hatching, the worms are placed under a layer of gauze, on which is spread a layer of finely chopped mulberry leaves. For six weeks, the worms eat almost continuously. At the end of this period, they are ready to spin their cocoons, and branches of trees or shrubs are placed in their rearing houses. The worms climb these branches and make their cocoons in one continuous thread, taking about eight days for the process. The amount of usable silk in each cocoon is small, and about 5500 silkworms are required to produce 1 kg (2.2 lb) of raw silk.

After the complete cocoons have been gathered, the initial step in silk manufacture is to kill the insects inside them. Thus, the cocoons are first boiled or treated in ovens, killing the insects by heat.

The silk fibre is obtained from the cocoons by a delicate process known as reeling, or filature. The cocoons are first heated in boiling water to dissolve the gummy substance that holds the cocoon filament in place. After this heating, the filaments from four to eight cocoons are joined and twisted and are then combined with a number of other similarly twisted filaments to make a thread that is wound on a reel.

When each cocoon is unwound, it is replaced with another cocoon. The resulting thread, called raw silk, consists usually of 48 individual silk fibres. The thread is continuous and, unlike the threads spun from other natural fibres such as cotton and wool, is made up of extremely long fibres. Along with cocoons damaged by emerging worms used for breeding stock, the filaments from the coarse outer portion of the cocoon, which is removed by brushing before reeling, and the inner portion of the cocoon, which remains after reeling the raw silk, are mixed to produce a low grade of silk staple that is spun into yarn.

The next step in the processing of silk is the twisting of one or more threads of the raw silk into a strand sufficiently strong for weaving or knitting. This procedure is called throwing. Four different types of silk thread may thus be produced: organzine, crepe, tram, and thrown singles. Organzine is a thread made by giving the raw-silk thread a preliminary twist in one direction and then twisting two of these threads together in the opposite direction at the rate of about 4 turns/cm (10 turns/in). Crepe is similar to organzine but is twisted to a much greater extent, usually between 16 and 32 turns/cm (40 and 80 turns/in).

Tram is made by twisting in only one direction two or more raw-silk threads, with 8 to 12 turns/cm (20 to 30 turns/in). Thrown singles are individual raw-silk threads that are twisted in only one direction, the number of turns depending on the quality of thread desired. In general, organzine thread is used for the warp threads of materials, and tram threads for the weft, or filling. Crepe thread is employed in the weaving of characteristic crinkly fabrics, and single thread is used for sheer fabrics.

SILKWORM

Silkworm, common name for the silk-producing larvae of any of several species of moths. Silkworms possess a pair of specially modified salivary glands called silk glands, or sericteries, which are used in the production of cocoons. The silk glands secrete a clear, viscous fluid that is forced through openings, called spinnerets, on the mouthparts of the larva; the fluid hardens as it comes into contact with air. The diameter of the spinneret determines the thickness of the silk thread produced.

The best-known silkworm is the larvae of the common, domesticated silkworm moth. This moth, native to China, was introduced into Europe and western Asia in the 6th century AD and into North America in the 18th century. The moth has been cultivated for many centuries and is no longer known in the wild state. Breeders have produced many varieties of the moth, the most important of which produce three broods of young annually.

A typical adult silkworm moth is yellow or yellowish-white, with a thick, hairy body, and has a wingspread of about 3.8 cm (about 1.5 in). The adult has rudimentary mouthparts and does not eat during the short period of its mature existence; the female dies almost immediately after depositing the eggs, and the male lives only a short time thereafter. The female deposits 300 to 400 bluish eggs at a time; the eggs are fastened to a flat surface by a gummy substance secreted by the female. The larvae, which hatch in about ten days, are about 0.6 cm (about 0.25 in) long. The larvae feed on leaves of white mulberry, Osage orange, or lettuce. Silkworm caterpillars that are fed mulberry leaves produce the finest quality silk. Mature larvae are about 7.5 cm (about 3 in) long and yellowish-gray or dark gray in colour.

About six weeks after hatching, the common silkworm stops eating and spins its cocoon. The length of the individual fibre composing the cocoon varies from 300 to 900 m (1000 to 3000 ft). The silkworm pupates for about two weeks; if allowed to complete its pupation period, it emerges as an adult moth. Tearing during emergence damages the silken cocoon beyond commercial use. Therefore, in the commercial production of silk, only enough adult moths are allowed to emerge to ensure continuation of the species. Most of the silkworms are killed by heat, either by immersion in boiling water or by drying in ovens.

Other moths known as silkworm moths include the giant silkworm moths. The larvae of these large moths also spin silken cocoons, but they are less widely used for commercial silk production.

Scientific classification: Silkworms belong to the order Lepidoptera. The domesticated silkworm moth makes up the family Bombycidae and is classified as *Bombyx mori.* The giant silkworm moths make up the family Saturniidae.

WILD SILKS

The ravages of pebrine and other diseases had the effect of attracting prominent attention to the numerous other insects, allies of the mulberry silkworm, which spin serviceable cocoons. It had been previously pointed out by Captain Hutton. who devoted great attention to the silk question as it affects the East Indies, that at least six species of *Bombyx,* differing from *B. mori,* but also mulberry-feeding, are more or less domesticated in India. These include *B. textor,* the boropooloo of Bengal, a large species having one generation yearly and producing a soft flossy cocoon; the Chinese monthly worm, *B. sinensis,* having several generations, and making a small cocoon; and the Madrasi worm of Bengal *croesi*), the Dassee or Desi worm of Bengal (*B. fortunatus*) and *arracanensis,* the Burmese worm - all of which yield several *Antheraea pernyi* (male).

Generations in the year and form reelable cocoons. Besides these there are many other mulberry-feeding *Bombycidae* in the East, principally belonging to the genera *Theophila* and *Ocinara,* the cocoons of which have not attracted cultivators.

The moths yielding wild silks which have obtained most attention belong to the extensive and handsome family *Saturnidee.* The most important of the species at the present time is the Chinese tussur or tasar worm, *Antheraea pernyi* an oakfeeding species, native of Mongolia, from which is derived the greater part of the so-called tussur silk now imported intoEurope.

Closely allied to this is the Indian tussur moth. *Antheraea mylitta,* found throughout the whole of India feeding on the bher-tree, *Zizyphus jujuba,* and on many other plants. It yields a large compact cocoon of a silvery grey colour, which Sir Thomas Wardle of Leek, who devoted a great amount of attention to the wild-silk question, succeeded in reeling.

Next in promising qualities is the muga or moonga worm of Assam, *Antheraea assaina,* a species to some extent domesticated in its native country. Its eggs were first sent to Europe by Duchene du Bellecourt, French consulgeneral in Japan in 1861; but early in March following they hatched out, when no leaves on which the larvae would feed were to be found. In April a single worm got oak-buds, on which it throve, and ultimately spun a cocoon whence a female moth issued, from which Guerin Meneville named and described the species. A further supply of eggs was secretly obtained by a Dutch physician Pompe van Meedervoort in 1863, and, as it was now known that the worm was an oak-feeder, and would thrive on the leaves of European oaks, great results were anticipated from the cultivation of the yama-mai.

These expectations, however, for various reasons, have been disappointed. The moths hatch out at a period when oak leaves are not ready for their feeding, and the silk is by no means of a quality to compare with that of the common mulberry worm. The mezankoorie moth of the Assamese, *Antheraea mezankooria,* yields a valuable cocoon, as does also the Atlas moth, *Attacus atlas,* which has an omnivorous larva found throughout India, Ceylon, Burmah,

China and Java. The Cynthia moth, *Attacus cynthia,* is domesticated as a source of silk in certain provinces of China, where it feeds on the *Ailanthus glandulosa.*

The eria or arrindi moth of Bengal and Assam, *Attacus ricini,* which feeds on the castor-oil plant, yields seven generations yearly, forming loose flossy orange-red and sometimes white cocoons. The ailanthus silkworm of Europe is a hybrid between *A. cynthia* and *A. ricini,* first obtained by Guerin Meneville, and now spread through many silk-growing regions. These are only a few of the moths from which silks of various usefulness can be produced; but none of these presents qualities, saving perhaps cheapness alone, which can put them in competition with common silk.

Physical and Chemical Relations of Silk

Common cocoons enclosing chrysalides weigh each from 16 to 50 grains, or say from 300 to 600 of small breeds and from 270 to 300 of large breeds to the lb. About one-sixth of this weight is pure cocoon, and of that one-half is obtainable as reeled silk, the remainder consisting of surface floss or blaze and of hard gummy husk. As the outer flossy threads and the inner vests are not reelable, it is difficult to estimate the total length of thread produced by the silkworm, but the portion reeled varies in length and thickness, according to the condition and robustness of the cocoon, in some breeds giving a result as low as Soo metres, and in others 900 to 1200 metres. Under favourable conditions it is estimated that i r kilogrammes of fresh cocoons give 1 kilogramme of raw silk for commerce, and about the same quantity for waste spinning purposes.

As this estimation presents some difficulties and divergences, the size of the thread is generally defined commercially by deniers or decigrammes, those of the Anthereas (wild silks) being said to range from 5 to 8 deniers or decigrammes, results confirmed by actual experience with the reeled thread.

The silk of the various species of *Antheraea* and *Attacus* is also thicker and stronger at the centre of the reeled portion than towards its extremities; but the diameter is much greater than that of common silk, and the filaments under the microscope present the appearance of flat bands, the exudation from the two spinnerets being joined at their flat edges. On this account the fibres of tussur or tussore silk tend to split up into fine fibrillae under the various preparatory processes in manufacturing, and its riband structure is the cause of the glassy lustre peculiar to the woven and finished fibres.

Silk fibre consists essentially of a centre or core of fibroin, with a covering of sericin or silk albumen, and a little waxy and colouring matter. Fibroin, which is analogous to horn, hair and like dermal products, constitutes about 75 to 82% of the entire mass, and has a composition represented by the formula C15H23 506. It has the characteristic appearance of pure silk - a brilliant soft white body with a pearly lustre - insoluble in water, alcohol and ether, but it dissolves freely in concentrated alkaline solutions, mineral acids, strong acetic acid and in ammoniacal solution of oxide of copper. Sericin, which constitutes the gummy covering (Fr. *gres*) of the fibre, is a gelatinous body which dissolves readily in warm soapy solutions, and in hot water, in which on cooling it forms a jelly with even as little as 1% of the substance.

It is precipitated from hot solutions by alcohol, falling as a white powder. Its formula is C15H25N508. According to P. Bolley, the glands of the silkworm contain semi-liquid fibroin alone, and it is on exposure to the air that Cocoon of *Antheraea pernyi.*

The surface is acted on by oxygen, transforming the external pellicle into the more soluble form of sericin. Silk is highly hygroscopic. If desiccated at 250° F. it will be found to lose from 10 to 15% of moisture according to the condition of the silk. It is a most perfect non-conductor of electricity, and in its dry state the fibres frequently get so electrically excited as to seriously interfere with their working, so that it becomes necessary to moisten them with glycerin or soapy solutions. Silk is readily distinguished from wool and other animal fibres by the action of an alkaline solution of oxide of lead, which darkens wool, &c., owing to the sulphur they contain, but does not affect silk, which is free from that body. Again, silk dissolves freely in common nitric acid, which is not the case with wool.

From vegetable fibres silk is readily distinguished by the bright yellow colour it takes from a solution of picric acid, which does not adhere to vegetable substances. The rod-like appearance of silk and its absence of markings under the microscope are also easily recognizable features of the fibre.

Silk Manufacture. Here we must distinguish between the reeled silk and the spun or waste silk manufactures. The former embraces a range of operations peculiar to silk, dealing as they do with continuous fibres of great length, whereas in the spun silk industry the raw materials are treated by methods analogous to those followed in the treatment of other fibres. It is only floss, injured and unreelable cocoons, the husks of reeled cocoons, and other waste from reeling, with certain wild silks, which are treated by the spun silk process, and the silk thereby produced loses much of the beauty, strength and brilliance which are characteristic of the manufactures from reeled silk.

Filature or Reeling

When the cocoons have been gathered the chrysalides they contain are killed either by dry heat or by exposure to steam. All cocoons stained by the premature death of the chrysalides (*chiques*), pierced cocoons, and any from other causes rendered unreelable, are put aside for the spun-silk manufacture. Then the uninjured cocoons are by themselves sorted into classes having similar shades of colour, size and quality of fibre. This assortment is of great consequence for the success of the reeling operations, as uniformity of quality and evenness and regularity of fibre are the most valuable features in raw silk. The object of reeling is to bring together the filaments (*bave*) from two or more (generally four or five, but sometimes up to twenty) cocoons, and to form them into one continuous, uniform, and regular strand, which constitutes the " raw silk " of commerce.

To do this, the natural gum of the cocoons which holds the filaments together must be softened, the ends of the filaments of the required number of cocoons must be caught, and means must be taken to unwind and lay these filaments together, so as to form a single uniform rounded strand of raw silk. As the reeling proceeds the reeler has to give the most careful attention to the thickness of the strand being produced, and to introduce new cocoons in place of any from which the reelable silk has become exhausted. In this way a continuous uniform fibre or strand of raw silk of indefinite length is produced.

The apparatus used for these purposes in some localities is of a very primitive kind, and the reeling being uneven and lumpy the silk is of inferior quality and low value. With comparatively simple appliances, on the other hand, a skilled reeler, with trained eye and delicate touch, can produce raw silk of remarkably smooth and even quality. According to the method commonly adopted in North Italy and France the cocoons are for a few minutes immersed in water a little under the boiling point, to which a small quantity of alkali has been added. A girl with a small hand brush of twigs keeps stirring them in the water till the silk softens, and the outer loose fibres (floss) get entangled with the twigs and come off till the end of the main filament (*maitre brin*) is found.

These ends being secured, the cocoons are transferred to a basin or tray containing water heated to from 140° to 150° F., in which they float while the silk is being reeled off. If the water is too cold the gum does not soften enough and the cocoons rise out of the basin in reeling; if it is too hot the cocoons collapse and fall to the bottom. The ends of the requisite number of filaments being brought together, they are passed through an eyelet or guide, and similarly another equal set are passed through a corresponding guide.

The two sets of filaments are then crossed or twisted around each other several turns as if to make one thread, after which they are separated and passed through separate guides to the reel round which they are separately wound. When a large number of cocoons are to be combined into one strand they may be reeled from the tray in four sets, which are first crossed in pairs, then combined into two, and those two then crossed and afterwards combined into a single strand. The object of crossing (*croissage*) is to round, smooth and condense the separate filaments of each set into one strand, and as the surface of the filaments is gummy and adhesive it is found on drying that they have agglutinated into a compact single fibre of raw silk. In the most approved modern filatures there is a separate cocoon boiler (*cuiseuse*), an oblong tank containing water boiled by steam heat.

In these the cocoons are immersed in rectangular perforated boxes for about three minutes, when they are transferred to the beating machine (*batteuse*), an earthenware trough having a perforated false bottom through which steam keeps the water at a temperature of from 140° to 160°. In this water the cocoons are kept stirring by small brushes rotated by mechanical means, and as the silk softens the brushes gradually rise out of the water, bringing entangled with them the loose floss, and thereby revealing the main filament of each cocoon.

The cocoons are next, in sufficient number, transferred to the reeler's tray (*bacinella*), where the water is heated to about 140° to 150°. From the tray the filaments are carried through a series of porcelain and glass eyelets, so arranged that the strand returns on itself, two portions of the same strand being crossed or intertwisted for rounding and consolidation, instead of the *croissage* of two separate strands as in the old method. The reel to which the raw silk is led consists of a light six-armed frame, enclosed within a wooden casing having a glass frame in front, the enclosure being heated with steam-pipes.

To keep the strands from directly overlaying each other and so adhering, the last guide through which the silk passes has a reciprocating motion whereby the fibre is distributed within certain limits over the reel. 12 presents a sectional view of a reeling apparatus as used in Italy, and shows the passage of the thread from the basin to the reel, the threads being twisted around by the tavelette to give roundness to the thread, but though the principle remains much the same, great improvements have been made on this model.

THROWING

Raw silk, being still too fine and delicate for ordinary use, next undergoes a series of operations called throwing, the object of which is to twist and double it into more substantial yarn. The first operation of the silk throwster is winding. He receives the raw silk in hanks as it is taken from the reel of the filature, and putting it on a light reel of a similar construction, called the swifts. The winds it on bobbins with a rapid reciprocating motion, so as to lay the fibre in diagonal lines.

These bobbins are then in general taken to the first spinning frame, and there the single strands receive their first twist, which rounds them, and prevents the compound fibre from splitting up and separating when, by the subsequent scouring operations, the gum is removed which presently binds them into one. Next follows the operation of cleaning, in which the silk is simply reeled from one bobbin to another, but on its way it passes through a slit which is sufficiently wide to pass the filament but stops the motion when a thick lump or nib is presented.

In the doubling, which is the next process, two or more filaments are wound together side by side on the same reel, preparatory to their being twisted or thrown into one yarn. Bobbins to the number of strands which are to be twisted into one are mounted in a creel on the doubling frame, and the strands are passed over smooth rods of glass or metal through a reciprocating guide to the bobbin on which they are wound. Each separate strand passes through the eye of a faller, which, should the fibre break, falls down and instantly stops the machine, thus effectually calling attention to the fact that a thread has failed. The spinning or throwing which follows is done on a frame with upright spindles and flyers, the yarn as it is twisted being drawn forward through guides and wound on revolving bobbins with a reciprocating motion. From these bobbins the silk is reeled into hanks of definite length for the market. Numerous attempts have been made to simplify the silk-throwing by combining two or more operations on one machine, but not as yet with much success.

According to the qualities of raw silk used and the throwing operations undergone the principal classes of thrown silk are - (1) " singles," which consist of a single strand of twisted raw silk made up of the filaments of eight to ten cocoons; (2) tram or weft thread, consisting of two or three strands of raw silk not twisted before doubling and only lightly spun (this is soft, flossy and comparatively weak); (3) organzine, the thread used for warps, made from two and rarely three twisted strands spun in the direction contrary to that in which they are separately twisted. Silks for sewing and embroidery belong to a different class from those intended for weaving, and thread-makers throw their raw silks in a manner peculiar to themselves.

NUMBERING OF SILK

The metric system of weights and measures has been adopted so widely that it forms the most suitable basis for the *titrage* or counts of yarns. The permanent committee of the Paris International Congress of 1900, which was held for the purpose of unification of the numerotage of counts, unanimously decided - (a) With reference to cotton, silk and other textiles spun from fibres, that they should be based on a fixed weight and variable length, the unit being one metre to one gramme. Thus number loo would be 100 metres per gramme calculated on the single strand. (*b*) With reference to raw and thrown silk, in order to enable the count to show the degrees of variation incidental to this class of material, it was decided for a basis of a fixed length and variable count weight. The length of skein adopted was 450 metres and the unit of length the half decigramme. Thus the count of silk is expressed by the number of half decigrammes which the length of 450 metres weighs. This obtains whether in the single, double or more threads joined together in the doubling.

This latter differs very little in actual practice from the previous method of determination by the number of deniers per 476 metres, the denier being calculated on the equivalent of 0.0531 gramme, the English equivalent showing 333 deniers per one dram avoirdupois.

As the old systems of counts have some technical conveniences they will no doubt be retained for some time. In some districts, especially in Yorkshire, the count is based on the number of yards per ounce, and in others the older method of drams avoirdupois per woo yard skein. The English cotton yarn and spun silk counts are reckoned upon the number of hanks of 840 yds. in 'lb of silk, cotton being reckoned upon the single thread and spun silk on the doubled or finished thread. Thus 2/40 9 cotton indicates single 40 9 doubled to 20 hanks by 840 yds. to the lb., while 40/2 fold spun silk means a single 80 9 doubled to give 40 hanks of 840 yds. to the lb. All continental conditioning establishments now formulate their tests for counts on the agreement arrived at by the International Congress of 1900.

Conditioning

Silk in the raw and thrown state absorbs a large amount of moisture, and may contain a percentage of water without being manifestly damp. As it is largely sold by weight it becomes necessary to ascertain its condition in respect of absorbed water, and for that purpose official conditioning houses are established in all the considerable centres of silk trade. In these the silk is tested or conditioned, and a certificate of weight issued in accordance with the results. The silk is for four hours exposed to a dry heat of 230° F., and immediately thereafter weighed. To the weight 11% is added as the normal proportion of water held by the fibre.

Scouring

Up to this point the silk fibre continues to be comparatively lustreless, stiff and harsh, from the coating of albuminous matter (gum or *gres*) on its surface. As a preliminary to most subsequent processes the removal of the whole or some portion of this gum is necessary by boiling-off, scouring or *decreusage.* To boil off say 300 lb of thrown silk, about 60 lb of fine white soap is shred, and dissolved in about 200 gallons of pure water. This solution is maintained at a heat of 195°, and in it the hanks of raw silk are immersed, hung on a wooden rod, the hanks being continually turned round so as to expose all portions equally to the solvent influence of the hot solution.

After being dried, the hanks are packed in linen bags and boiled for three hours in a weaker soapy solution, then washed out in pure warm water and dried in a centrifugal hydroextractor. According to the amount of gum to be boiled off the soap solutions are made strong or weak; but care has to be exercised not to overdo the scouring, whereby loss of strength, substance and lustre would result. For some purposes-making of gauzes, crapes, flour-bolting cloth and for what is termed "souples"-the silk is not scoured, and for silks to be dyed certain dark colours half-scouring is practised. The perfect scouring of silks removes from 20 to 27 of their weight, according to the character of the silk and the amount of soap or oil used in the working. Scouring renders all common silks, whether white or yellow in the raw, a brilliant pearly white, with a delicate soft flossy texture, from the fact that the fibres which were agglutinated in reeling, being now degummed, are separated from each other and show their individual tenuity in the yarn.

Silks to be finished white are at this point bleached by exposure in a closed chamber to the fumes of sulphurous acid, and at the close of the process the hanks are washed in pure cold water to remove all traces of the acid.

SILK WEIGHTING

Into the dyeing of silk it is not here necessary to enter, except in so far as concerns a nefarious practice, carried on in dye-houses, which has exercised a most detrimental influence on the silk trade. Silk, we have seen, loses about one-fourth of its weight in scouring.

To obviate that loss it has long been the practice to dye some dark silks " in the gum," the dye combining in these cases with the gum or gelatinous coating, and such silks are known as " souples. Both in the gum and in the boiled-off state silk has the peculiar property of imbibing certain metallic salts largely and combining very firmly with them, the fibre remaining to external appearance undiminished in strength and lustre, but much added to in size and weight. Silk in the gum, it is found, absorbs these salts more freely than boiled-off; so to use it for weighting there are these great inducements - a saving of the costly and tedious boiling-off, a saving of the 25% weight which would have disappeared in boiling and a surface on which much greater sophistication can be practised than on scoured silk.

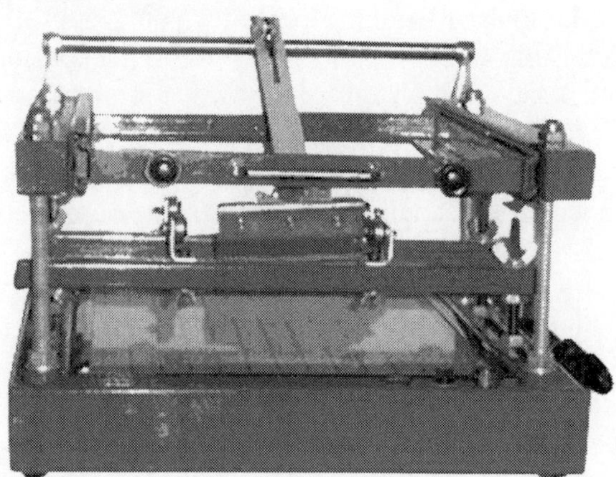

Fig. Silk Weighting Machine

In dyeing a silk black a certain amount of weight must be added; and the common practice in former times was to make up on the silk what was lost in the scouring. Up to 1857 the utmost the dyer could add was "weight for weight," but an accidental discovery that year put dyers into the way of using tin salts in weighting with the result that they were enabled to add 40 oz. to scoured silk, 120 oz. to souples and as much as 150 oz. to spun silks.

This excessive adulteration quickly worked its own cure by a decreased consumption, and the weighting in practice in 1910 is confined to moderate and safer limits. The use of tin salts, especially stannic chloride, SnC1 4, enables dyers to weight all colours the same as black. In his " Report on English Silk Industry " to the Royal Commission on Technical Instruction (1885) Sir Thomas Wardle of Leek says: "Colours and white of all possible shades can very easily be imparted to this compound of silk and tin, and this method is becoming extensively used in Lyons.

Thus weighting, which was until recently thought to apply only to black silks, and from which coloured silks were comparatively free, is now cheapening and deteriorating the latter in pretty much the same ratio as the former. Thus the protoand per-salts of iron, as well as the protoand per-salts of tin, including also a large variety of tannin, sumac, divi-divi, chestnut, valonia, the acacias (*Areca Catechu* and *Acacia Catechu* from India), from which are obtained cutch and gambier, &c., are no longer used solely as mordants or tinctorial matters, but mainly to serve the object of converting the silk into a greatly-expanded fibre, consisting of a conglomeration of more or less of these substances." Sugar also is employed to weight silk.

On this adulterant Sir Thomas Wardle remarks "With a solution of sugar, silk can have its weight augmented from I oz. to 3 oz. per lb. This method of weighting was not first used by the throwsters, as sugar is known to have been used for adulterating and loading gum silk for a very long time, and then the idea was afterwards applied to silk after the dyeing operations. It is much resorted to for weighting coloured silks by dyers on the continent, and, though a very clumsy method, no substitute has been found so cheap and easy of application. Bichloride of tin, having chemical affinity for silk fibre, bids fair to extinguish the use of sugar, which, from its hygrometric qualities, has a tendency to ruin the silk to which it is applied, if great care be not taken to regulate the quantity.

There is not the slightest use or excuse for the application of sugar, except to cheapen the silk by about 15 to 20 *Wild Silk* Dyeing. Among the disadvantages under which the silks of the wild moths long laboured one of the most serious was the natural colour of the silks, and the extreme difficulty with which they took on dyes, specially the light and brilliant colours. For success in coping with this difficulty, as well as in dealing with the whole question of the cultivation and employment of wild silks, the unwearying patience and great skill of Sir Thomas Wardle of Leek deserve special mention here.

The natural colour of tussur silk is a greyish fawn, and that shade it was found impossible to discharge by any of the ordinary bleaching agents, so as to obtain a basis for light and delicate dyes. Moreover, the chemical character of the tussur silk differs from that of the mulberry silk, and the fibre has much less affinity for tinctorial substances, which it takes up unevenly, requiring a large amount of dye-stuffs. After protracted experimenting Sir Thomas Wardle was able in 1873 to show a series of tussurs well dyed in all the darker shades of colour, but the lighter and bright blues, pinks, scarlets, & c., he could not produce, Subsequently Tessie du Motay found that the fawn colour of natural tussur could be discharged by solution of permanganate of potash, but the oxidizing action was so rapid and violent that it destroyed the fibre itself.

Gentler means of oxidation have since been found for bleaching tussur to a fairly pale ground. The silk of the eria or castor-oil worm (*Attacus ricini*) presents the same difficulties in dyeing as the common tussur. A portion of the eria cocoons are white, while the others are of a lively brown colour, and for the dyeing of light colours the latter require to undergo a bleaching process. The silk takes up colour with difficulty from a strong vat, and is consequently costly to dye.

Moonga silk from *Antheraea assama* has generally a rather darkbrown colour, but that appears to be much influenced by the leaves on which the worm feeds, the cocoons obtained on the champaca tree (*Michelia champaca*) giving a fine white fibre much valued in Assam. The dark colours are very difficult to bleach, but the silk itself takes dye-colours much more freely and evenly than either tussur or eria silk. (F. W. *) *Trade and Commerce.* About the beginning of the 19th century the chief silk-producing regions of the world were the Levant (including Broussa, Syria and Persia), India, Italy and France, the two first named sending the low-priced silk, the other two the fine qualities.

Between 1840 and 1850, after the opening of trade with China, large quantities of silk were sent from the northern port of Shanghai, and afterwards also from the southern port of "Canton. The export became important just at the time when disease in Europe had lessened the production on the continent.

This increased production of medium silk, and the growing demand for fine sorts, induced many of the cocoon-growers in the Levant to sell their cocoons to Europeans, who reeled them in Italian fashion under the name of " Patent Brutia," thus producing a very fine valuable silk. In 1857 commenced the exportation of Japan silk, which became so fierce a competitor with Bengal silk as gradually to displace it in favour; and the native silk reeled in Bengal has almost ceased to be made, only the best European filatures, produced under the supervision of skilled Europeans, now coming forward.

China and Japan, both of which contribute so largely to the supplies that appear in European and American statistics, only export their excess growth, silk-weaving being carried on and native silk worn to an enormous extent in both countries. The other Asiatic exporting countries also maintain native silk manufactures which absorb no inconsiderable proportion of their raw material.

Since about 1880 the silk production of the world (including only exports from the East) has more than doubled, the variations owing to partial failures from some countries being more than compensated by increase from others. The supplies available for European and American consumption have been carefully tabulated by the Lyons Chamber of Commerce, as shown by the table.

While the tables indicate the fluctuations of supply they show generally that Asiatic countries, in addition to supplying the necessities for their home trade, export to Europe and America about threefifths of the whole of the silk consumed in Western manufactures.

Up to the year 1860 the bulk of the silks from the East was shipped to London, but subsequently, owing to the importance of continental demands, a large portion of the supplies has been unshipped at Genoa and Marseilles (especially the finer reeled silks from Japan and Canton), which are sold in the Milan and Lyons markets.

CATERPILLAR

Caterpillar, larval stage of butterflies and moths, members of the order Lepidoptera, and corresponding in this special order to the grub, maggot, or larva phase in the life history of other insects. The caterpillar develops like any other larva from the segmented egg and differentiating embryo and undergoes several moltings, or ecdyses. It later falls into a quiescent pupa stage, and the pupa is usually sheathed in a silken cocoon. It may be fixed or free, suspended by one thread or more to a leaf or branch, or hidden underground.

Comparatively few caterpillars reach maturity. Many are destroyed by weather; by birds, reptiles, and other animals; and by insects such as ichneumon wasps, tachinid flies, and ground beetles. Ichneumon wasps pierce the caterpillars and make them receptacles for their eggs and, later, edible cradles for their larvae.

SPIDER (ARTHROPOD)

Spider (arthropod), any of a large group of *invertebrates* (animals without backbones) that have spinning glands used to produce silken threads and webs. There are about 40,000 species of spiders. Spiders are found worldwide, except for in the oceans, and they live in all habitats and at most elevations.

Spiders, along with daddy longlegs, mites, ticks, and scorpions, belong to a group of animals known as arachnids. Spiders are not insects. Spiders have a different body structure than insects.

The bodies of spiders divide into two parts (a fused head and thorax and an abdomen), and spiders have eight legs and eight eyes; they lack antennae and wings. In contrast, the bodies of insects form three parts (head, thorax, and abdomen), and insects have six legs, two eyes, two antennae, and, typically, four wings.

All spiders are *carnivorous* (feeding mainly on the flesh of other animals). Spiders eat insects and sometimes other arthropods (invertebrate animals with jointed limbs, segmented bodies, and hard shells known as exoskeletons), including other spiders. Scientists often divide spiders into two types: web spiders and ground spiders. Web spiders produce webs to capture prey, while ground spiders hunt prey directly without using a web. Almost all spiders use poison glands to kill or paralyze their prey or to defend themselves.

Contrary to popular belief, most spider bites are not dangerous to humans. Of the 40,000 species of spiders, only about 30 species produce bites that may cause illness. Spiders rarely attack humans unless they feel threatened, and if they do bite, the wound is rarely serious. There is absolutely no reason to kill any spider or to call an exterminator if you have spiders in your house. As an old English saying goes, "If you want to live and thrive, let a spider run alive!"

EXTERNAL BODY

Spiders range in size from less than 1.0 mm (0.04 in) to more than 10 cm (4 in) in length, with a leg span of up to 20 cm (8 in). A spider's body is divided into two parts: the front portion, called the prosoma or cephalothorax, and the rear portion, called the opisthosoma or abdomen. A narrow stalk called the pedicel connects these two parts.

A hard shell called an exoskeleton covers the entire body of a spider. The exoskeleton is made of cuticle, a material composed of a combination of protein and tough fibres called chitin. The cuticle forms thin layers stacked on top of one another, an arrangement that improves the strength and elasticity of the exoskeleton. The spider's cuticle provides attachment sites for many muscles, and it also prevents *desiccation* (loss of body water). The cephalothorax cuticle is strong and stiff, while the cuticle of the abdomen is soft and extensible. As a spider grows, it sheds or molts its exoskeleton and grows a new one to cover its larger body.

Cephalothorax Structures

The cephalothorax contains a number of structures and appendages: one pair of biting mouthparts known as chelicerae; a pair of poison glands; one pair of short, leglike appendages called pedipalps or palps; and four pairs of legs. The spider's eight eyes are also located on the cephalothorax.

Mouthparts

When a spider catches prey it uses a pair of jointed appendages known as the chelicerae, located in front of the mouth opening. Chelicerae resemble tiny pocketknives. Each chelicera has a sharp fang that swings out of its resting position to stab into the victim. Near the tip of the fang is a duct opening that comes from a poison gland. The fang acts like a hypodermic needle—it ejects venom from the poison gland and delivers it into the prey. Spiders also use chelicerae as multipurpose tools. They have been called the "hands" of the spider. Spiders can use their chelicerae to perform tasks such as digging burrows in the soil and transporting small prey.

Poison Glands

Most spiders have a pair of poison glands that lie within the cephalothorax. Each bulblike poison gland produces and stores toxin. A muscle spirals around the gland. When this muscle contracts, it squeezes poison from the gland through a duct into the fangs of the chelicerae, which then pass the poison into the prey.

Palps and Legs

Behind the chelicerae are a pair of palps, segmented limbs that are used in feeding and as feelers. Male spiders also use palps to transfer sperm to females during mating. Adjacent to the palps are four pairs of long, hairy legs. Unlike human hair, each spider hair found on the legs acts as a sensory organ, sensitive to touch and vibration.

Each leg is made up of seven jointed segments, called the coxa, trochanter, femur, patella, tibia, metatarsus, and tarsus. More than 30 muscles control the movement of each leg. In addition, some joints of the leg move by the hydraulic action of body fluid.

The tips of the legs have two or three small claws that are used for climbing or grasping the spider's silk thread. Many ground spiders have specialized adhesive hairs beneath their claws, known as claw tufts or scopulae. These claw tufts enable the spiders to walk surefootedly on smooth, vertical surfaces—even upside down on glass.

Sensory Organs

Most spiders are active at night, and as a result they use their other senses more than they use their eyesight, which is not well developed. In addition to the thousands of hairs found on the palps and legs that are highly sensitive to touch and vibrations, spiders also have hairs on their feet that they use to taste things.

Most spiders have four pairs of simple eyes (eyes with a single lens) that are located on the front of the cephalothorax. The eyes are usually grouped into two or three rows that form specific patterns in different spider families. This eye arrangement is often used to identify and classify a spider.

Unlike spiders that are active at night, spiders that are active during the day, such as jumping spiders and lynx spiders, typically have good vision at close range (around 10 to 20 cm, or 4 to 8 in). Their vision easily rivals the eyesight of many insects, which have *compound eyes* (eyes with multiple lenses).

Spider Abdomen

The spider's abdomen is soft and saclike. On the underside of the tip of the abdomen are three pairs of spinnerets. Each spinneret is studded with many fine, hairlike tubes called spigots, which produce a variety of silk threads. The spigots lead to several large silk glands inside the abdomen. Silk is formed as a liquid inside these abdominal glands.

As the silk is drawn out through the spigots, protein molecules within the silk line up parallel to one another, causing the silk to harden and form strong, elastic filaments. The hardening of silk results from the drawing-out process through the spigots, not from exposure to air, as is commonly believed. Several silk threads produced by different spigots may fuse to form a stronger one. Spinnerets are actually shortened limbs. They can move to place silk strands in precise locations when the spider builds a web or wraps prey in silk.

INTERNAL ANATOMY AND FUNCTION

The cephalothorax houses part of the digestive system and the central nervous system. The abdomen contains most of the spider's vital organs, including a long, tubular heart; respiratory organs; reproductive organs; and excretory organs.

Breathing

Spiders use two types of breathing organs: book lungs and tracheal tubes. Narrow slits on the underside of the abdomen lead to two or four respiratory organs called book lungs. These organs are so named because they consist of alternating layers of air spaces and thin leaflets of cuticle filled with blood, making the structures resemble a slightly opened book. Oxygen from the air passes through the extremely thin cuticle leaflets directly into the blood.

A small opening in front of the spinnerets, known as a spiracle, leads to tracheal tubes made of cuticle. These tiny tubes branch and spread throughout the body. Air enters the spiracle and passes through the tracheal tubes so that oxygen can travel to all the spider's body tissues.

Blood Circulation

Spider blood, also known as hemolymph, contains many blood cells with oxygen-carrying pigments called hemocyanin, which give the blood a light blue colour. In contrast, the primary components of human blood are red blood cells carrying the red pigment hemoglobin. Spider blood also contains many other types of blood cells that play a role comparable to that of the white blood cells of humans. Among other functions, these cells play a role in blood clotting after an injury.

The spider's long, tubular heart lies toward the back side of the abdomen. When the heart contracts, it pumps blood forward into the cephalothorax and backward into the abdomen. Blood travels through closed tubes, or arteries, into spaces in the body cavity. From these spaces the blood travels to the book lungs, where it releases carbon dioxide and picks up a fresh supply of oxygen before returning to the heart.

Digestion

The digestive system consists of a branched tube that extends from the mouth to the anus. In the cephalothorax, the tube enlarges to form a stomach with powerful muscles. When these muscles contract, they produce a powerful sucking action that pulls food into the midgut.

Spider digestion is unusual in that it begins outside of the spider's body. When a spider captures an insect or other animal, it uses its chelicerae to pierce the prey and inject poison into the wound to paralyze or kill the animal. The spider then vomits juices containing digestive enzymes into the wound of the victim to break down and liquefy its body tissue. This liquefied tissue is then drawn through the spider's mouth and into its body by the sucking action of the stomach. Two mechanical filters in the mouth prevent solid food particles from passing into the digestive system.

From the stomach, food passes into the midgut, which branches throughout the entire body. Enzymes secreted by the midgut further break down the liquefied food into nutrient molecules small enough to pass through the walls of the midgut into the blood. Nutrients can be stored for a long time in the spider's extensive digestive system, enabling many spiders to go for weeks or even months without the need to catch any prey.

Nervous System

Most arthropods have a central nervous system made up of a long chain of nerve cell centres called ganglia that run throughout the body. In spiders the ganglia are concentrated in the cephalothorax, where they condense into two compact masses: the sub-esophageal ganglion and the supra-esophageal ganglion. The sub-esophageal ganglion directs spider locomotion. The supra-esophageal ganglion is considered the brain of the spider. Sense organs throughout the body send information to this nerve centre, where information processes and complex functions begin. A spider's brain is relatively highly developed, enabling spiders to easily adapt to changes in their environment. Some scientists believe spiders can learn, and some have observed that spiders can remember where in their web they have stored captured prey; if the prey is removed, the spiders will continue searching for it in the same place for hours.

Reproduction

All species of spiders have two separate sexes, and the males are usually smaller than the females. The male spider has two sperm-producing testes. A sexually mature male spider uses its large palps to transfer sperm cells into the female during mating. In this process, the male builds a small, triangular sperm web onto which he deposits a drop of sperm from his abdomen. He then dips both palps into this droplet, drawing sperm cells into the palps as if by a tiny pipette.

The female reproductive system includes two egg-producing ovaries. After the male transfers sperm cells into the female's genital opening located on her abdomen, they are stored, sometimes for months, in tiny receptacles. These sperm cells fertilize the female's egg cells just before she deposits her eggs into a silky cocoon.

LIFE CYCLE

The life cycle of the spider consists of four stages: egg, larva, young spider, known as a nymph or spiderling, and adult. Like insects, spiders grow only by molting, a process that involves periodically shedding their exoskeleton. In each molting stage, young spiderlings resemble tiny adults, a process known as incomplete metamorphosis.

Courting and Mating

Spiders become sexually mature after their last molt, at which time females have developed functional ovaries and males have mature testes. In most spider species the male courts the female before mating occurs. After a male spider has filled its palps with sperm cells, he begins searching for a female. A male begins by identifying himself to a female so that she does not mistake him for potential prey. In some spiders, such as American tarantulas, this identification process involves the male repeatedly touching the female. More often, a male courting a female communicates with her over larger distances using vibrations. For instance, a male wolf spider uses its legs to drum on the ground. In some web spiders, the male attaches a special signal thread to the female's web. The male then drums or plucks the thread in a rhythm that indicates the vibration is caused by another spider of the same species and not by an ensnared insect. If a female is ready to accept a courting male, she may send signals back to him.

Locating the right female can be tricky for a male spider. Fortunately, female spiders produce certain chemical substances known as pheromones that aid spider courtship. A female may release these pheromones through the air (like a perfume) or she may deposit them on her silk threads. When a male spider encounters pheromones from a female of the same species, he becomes excited, even if the female is not present.

Spiders with better eyesight may rely mostly on visual signals during courtship. When a male notices a female, he starts a zigzag dance in front of her in which he raises his front legs, vibrates his palps, and twitches his abdomen. Each species uses a different courtship dance with unique movements. A female will only accept a male who performs a dance with movements specific to that species.

When a male finds an interested female, he inserts his sperm-containing palp into the female's genital opening. The process of mating can be very brief (a matter of seconds), or it can last several hours, depending on the species. In most species both sexes separate peacefully after mating. Contrary to popular belief, the female black widow spider does not kill her partner after mating. Depending on the species, a female may mate only once or she may mate with several males during her lifetime.

Eggs and Brood Care

In most species a female determines when sperm cells from the female's receptacles will fertilize her egg cells. Fertilization may occur a few weeks after mating, a strategy that enables the female to lay her eggs when she deems external conditions are best. The female then lays her eggs. Most spider species lay several hundred eggs; species of large spiders deposit several thousand within just a few minutes.

Typically the female lays her fertilized eggs in a silky case called a cocoon, which provides a protective and insulating environment for the developing spiders. Many females abandon their cocoons right after they deposit their eggs, although they may camouflage them or hang them in hidden locations. Other spiders guard and defend their cocoon until the eggs hatch.

Some spiders exhibit special brood care. Female wolf spiders attach their cocoons to their spinnerets and carry them around until the eggs hatch. The newly hatched spiderlings then climb onto their mother's back, where they stay for about a week before they leave to survive on their own. Occasionally young spiderlings stay in their mother's web for weeks. During this time the mother feeds her brood, transferring regurgitated food from her mouth to their mouths.

Development and Growth

Spider eggs hatch inside the cocoon. The first stage hatches from the egg as an immobile, milky white larva. Larvae are covered by an embryonic membrane and receive nourishment from yolk material within their abdomen. After one to two molts over about a two-week period, the larva changes into a mobile spiderling, also referred to as a nymph.

In order to grow to an adult size, spiderlings undergo a series of molts that enables them to increase in size. During molting the old cuticle slowly lifts off while a thin new cuticle forms underneath. The new cuticle is wrinkled and pliable at first, but as molting progresses and the spiderling grows, the new cuticle stretches to accommodate the larger spiderling body.

It later hardens into a new rigid exoskeleton that encases the larger spiderling. The number of molts between the spiderling and adult stage varies according to the size of the species. Small species may molt about 5 times, while some large tarantulas may molt as many as 40 times. For most spiders, a spiderling's last molt marks adulthood, when functional sexual organs have developed and growth halts. Some adult female spiders, such as American tarantulas, continue to molt repeatedly.

Most spiders live only 1 or 2 years. Notable exceptions are large female tarantulas, which can live up to 20 years. Male tarantulas live only 2 to 3 years. Many male spiders die soon after mating.

SILKS, THREADS, AND WEBS

Nearly all spiders produce silk composed of the protein fibroin. This is the same protein produced by silkworms, the larvae stage of certain silk moths. Each spider has four to eight different kinds of silk glands in its abdomen, and each gland produces a different type of silk with different properties. For instance, spiders lay out a line of dry silk behind them as they move about. This dragline acts as a safety line like that used by a mountain climber. Other glands produce cocoon threads that blanket and protect fertilized eggs. Still other glands produce sticky capture threads that ensnare prey.

Spider silk threads are very thin, about 1 micrometer (0.001 mm) in diameter. Each thread weighs very little. A spider web composed of 20 m (70 ft) of silk thread weighs less than 1/1000 of a gram. Despite its lightness, a spider silk thread is as strong as a nylon thread, but with more elasticity. The combination of strength and elasticity makes spider silk ideal for web building. Spider threads are tough enough to withstand the impact of a flying insect while being elastic enough not to tear apart with the captured prey's weight.

Types of Webs

Web patterns vary considerably, depending on the species of spider. Perhaps the most recognizable web is the almost circular orb web, in which an outer framework supports a continuous spiraling thread and a series of threads radiating from the centre of the web. Other web types may have a more irregular shape. Some spiders build irregular, flimsy webs. Common house spiders construct funnel webs, flat silk sheets with a raised tube in the corner that serves as a retreat for the spider. Cobweb spiders build an irregular silk meshwork with sticky threads at the bottom that trap insects. Sheet web spiders construct a horizontal silk sheet with a dome, from which the small spider hangs upside down.

Many spider webs are found near the ground or in low vegetation, although orb webs often span the open spaces between bushes or trees in order to trap flying insects. The size of a web depends on the size of the spider. Whether the web has a tight or wide mesh depends on the size of the prey the spider expects to capture.

Web Building

Web building is a complex process, but some spiders can complete a web in less than 30 minutes. Spiders typically build their webs at night, completely relying on their sense of touch, not eyesight, during construction. Each type of spider uses a different procedure when building a web. Spiders that weave orb webs generally begin by spinning a thread that is carried by air currents until it catches on a tree limb or other firm support. From this thread the spider lays down another thread to form a Y-shaped structure that is the basic framework of the web.

The spider then climbs to the midpoint of the Y-structure, known as the hub, and begins creating radius lines, or spokes, around the web. As the spider builds radius lines, it connects these lines with a few narrow circles of thread in the centre of the web that forms the auxiliary spiral. The auxiliary spiral prevents radius lines from sagging when the spider walks on them.

Using the auxiliary spiral as scaffolding, the spider begins the formation of the catching spiral, fastening sticky threads to each radius line. As the spider constructs the catching spiral, it dismantles the auxiliary spiral. The catching spiral extends from the periphery of the web and stops short of the web's hub. In the finished web, only remnants of the auxiliary spiral remain and there is an empty space in the web's centre known as the free zone.

The spider may sit upside down with its legs placed in the centre of its web to detect vibrations in the web when prey gets caught in the sticky catching spiral.

Others may hide nearby under a curled leaf and use the vibrations from a signal thread attached to the hub to stay informed when prey has struck the web. The orb web is built anew every day. Some species eat their old web before starting a new web, while others roll up the web and discard it as a tiny silk ball.

EVOLUTION

Spiders evolved around 400 million years ago. The ancient spiders that lived millions of years ago were relatively large, and their bodies were segmented. In contrast, almost all spiders living today have an unsegmented abdomen. Only members of the suborder Mesothelae still exhibit a segmented abdomen, and these spiders are generally considered the most primitive types of spiders.

The fossil record of spiders is rather limited. The oldest spider fossil was found in New York State in rocks dating back to the Devonian Period (about 410 million to 360 million years ago).

This fossil was remarkably well preserved. Using a microscope, scientists were able to recognize the spider's spinnerets and chelicerae. Some fossil spiders with segmented abdomens have been identified in rocks dating from the Carboniferous Period (360 million to 290 million years ago). Very few fossils have been found from the Mesozoic Era (240 million to 65 million years ago).

Spider webs rarely preserve as fossils, but scientists theorize that ancient spiders initially built irregular webs located near the ground and that webs located higher up in trees evolved only later. Orb webs, for instance, may be an adaptation for snaring insects flying higher up in the vegetation. Some scientists suggest that spiders may have influenced the evolution of insects—and vice versa. In an effort to escape ground-living spiders, insects may have evolved ways to fly.

At a later time, spiders may have evolved ways to develop aerial webs in order to catch flying insects.

TYPES OF SPIDERS

Spiders belong to the order Araneae. Scientists typically divide the 40,000 species of spiders into three suborders: the Mesothelae, the Mygalomorphae, and the Araneomorphae.

The suborder Mesothelae is a small, primitive group of spiders that have a segmented abdomen resembling that of early fossil spiders. Spiders of this suborder are sometimes called living fossils since they still exhibit primitive traits that no other living spiders share. Only 40 species from this suborder have been identified, and they live in Indonesia and Japan.

The suborder Mygalomorphae contains about 2,000 species, including tarantulas, trap-door spiders, and purse-web spiders. Although these spiders are not segmented, this group still retains some primitive characteristics, including four book lungs and chelicerae in which the biting parts lie parallel to each other and strike forward and down.

About 32,000 species, or almost 90 percent of all spiders, belong to the suborder Araneomorphae. These spiders are characterized by the presence of two book lungs and a tracheal tube breathing system.

They have chelicerae with opposing fangs that work together like a pair of pliers. Many araneomorphs are ground hunters. Although these spiders do not use a web to capture prey, they may use silk for other purposes such as for silk cocoons to protect eggs. The spiders highlighted are organized according to their hunting ability. Ground spiders hunt prey from the ground, while web spiders use webs to ensnare prey.

Tarantula

True tarantulas are a type of wolf spider that belong to the family Lycosidae in the suborder Araneomorphae. But most people use the term *tarantula* to refer to about 700 species of spiders belonging to the family Theraphosidae in the suborder Mygalomorphae. Sometimes known as American tarantulas, they are found in tropical regions throughout the world, with many species in the southwestern United States.

These giants of the spider world can attain a body length of 10 cm (4 in) with a leg span up to 20 cm (8 in), making it possible for some species to overpower small vertebrates (animals with backbones), such as frogs or lizards, for their meal. Their large body and long legs are covered with hairs, and they have powerful chelicerae, giving them a fearsome appearance.

Most tarantulas stay on the ground (often in burrows). As ground hunters, tarantulas are typical sit-and-wait predators—they wait for insects or small vertebrates to come near before they pounce and kill the prey with their strong chelicerae.

Perhaps as a result of their frightening size and appearance, tarantulas have gained a deadly reputation among humans. For the most part, however, tarantulas do not attack unless provoked, and their venom is not harmful to humans, although their strong chelicerae can cause painful wounds.

Tarantulas are popular pets, and some become so tame that they can be picked up and handled safely. Although the chance of receiving a bite is small, there is another danger: Many tarantulas brush off their abdominal hairs when they feel threatened. These barbed hairs fly through the air and can penetrate skin and the mucous membranes of the nose, causing a strong burning sensation.

TYPES OF SILK FABRIC

The finest silk fibres, and most of what we use today are produced by "cultivated" silkworms grown in a controlled environment.

The worms are fed a diet of mulberry leaves and increase their body size 10,000 times in their short life span. Once the cocoon is spun and before the worm hatches through the silk into a moth, the cocoon is soaked in hot water then unraveled, producing filaments that can be up to a mile long in size.

The raw silk is then processed to remove the sericin - the natural "gum" that protects the fibres and causes them to stick to each other as the cocoon was spun. Silk is a protein fibre, similar to wool or to human hair. It is only natural for silks to have some irregularities — sometimes called "slubs".

This is the nature of the 100% silk fabric. Silk that has been processed can be woven or knit into a variety of fabrics. Silk will shrink, so if you are using it to construct clothing, be sure to preshrink it.

Silk often has a wonderful feel and an almost iridescent sheen that makes us think of luxury. The weight of silk is shown as "mm" (pronounced "mommy") and varies within the different types of silk. The following list of various types of silk may help you understand some of the qualities of each.

Chiffon

Often the lightest weight and most diaphanous of the silks, Chiffon is also the most see-through. It creates the "billows" of fabric that add dimension to garments, but generally requires some kind of lining or backing unless it's used for scarves.

China Silk

China silk is a lightweight, sheer, plain-weave fabric. It's sometimes referred to as habutai, or habotai, or pongee. It is one of the less expensive and more commonly available silk fabrics.

Habotai can often be found as light as 5 mm and as heavy as 12 mm. Most of our scarves are made of 8 mm Habotai. When purchasing for clothing construction, or purchasing ready-made clothing, this fabric is not recommended for fitted garment styles because the seams will tear from the stress.

Crepe de Chine

Crepe de chine is a lightweight fabric made by twisting some fibres clockwise and others counterclockwise. The twisted fibres are then woven in a plain-weave fabric, but it's the twisted fibres, not the weave, that gives crepe its distinctive "pebbly" look and feel rather than a shiny luster. Both sides of the fabric look and feel the same. When purchasing ready-made clothing or considering this fabric for sewing, avoid using it in tailored styles because the fabric is too soft to hold a structured shape. Crepe de chine doesn't ravel as easily as other silk fabrics, but it will tear if not handled gently.

Charmeuse

Generally, when we think of traditional silk, this is the fabric we have in mind. The back of the fabric is a flattened crepe while the front is a shimmery satin weave. Charmeuse has even more drape than crepe de chine and works well for scarves, blouses and lingerie. Occasionally we will offer scarves in Charmeuse.

Jacquard

Jacquard silks offer various woven patterns, using matte and reflective threads to create a light and dark effect in the fabric. This effect is similar to brocade, although the Jacquard is originally created in one colour.

These are generally heavier weight and more densely woven. Patterns are often florals and paisleys. This added dimension (pattern) makes this fabric perfect for abstract for free-form dyeing.

Douppioni

Douppioni is a plain-weave fabric with slubbed ribs. It has a stiff, taffeta-like hand and is usually dyed in bright colours. Douppioni is often made into elegant evening gowns or semi-fitted vests and garments. But the style shouldnÕt be too fitted, because the fabric doesn't stand up well to stress and ravels easily.

It's often recommended that douppioni be dry cleaned to resist abrasions. However, as with most silk, you can generally wash douppioni with positive results. Just be sure to serge the raw edges first to prevent raveling. Washing will make the fabric lose some of its stiffness, which may be your preference, and the colour will soften as the excess dye is washed away.

Noil

Silk noil is made from the short fibres left after combing and carding so it doesn't shine like many other silk fabrics. Noil looks similar to cotton, but has the soft feel of silk against the skin. It also drapes better than cotton and resists wrinkling, so it's the perfect choice when traveling.It can be machine washed on gentle and dried on low, but this will cause a faded, "weathered" look. If you prefer bright colours, dry-clean or hand wash.

Raw Silk

Raw silk is any silk yarn or fabric that hasn't had the sericin—the natural "gum" that protects the fibre — removed. The fabric is stiff and dull and the sericin tends to attract dirt and odors.

Shantung

Once made from hand-reeled tussah silk, today's shantung is usually made with cultivated silk warp yarns and heavier douppioni filling yarns. Depending on the filling yarn, shantung may be lustrous or dull. It has a firm, semi-crisp hand and tends to ravel, so avoid close-fitting styles.

Tussah

Tussah silk, often called shantung, is made from the cocoons of wild tussah silk worms that eat oak and juniper leaves — their "natural" food. Because the worm isn't grown in a controlled environment, the moth hatches from the cocoon thus interrupting the filament length and making the fibres short and coarse instead of long and lustrous.

Tussah silk is difficult to dye and most often available in its natural colour, a creamy tan. Because of its irregular slubs and the fact that it ravels easily, tussah should be dry-cleaned. It is a good choice for traveling as it doesn't wrinkle easily.

CARING FOR SIL FABRIC

Silk is protein fibre, more similar to wool than to cotton. It is very similar to human hair. Remembering this will help when you think about how to wash or clean it. Silk is extremely strong, but repeated exposure to the sun will erode the fibre. As a result, silk fabrics are poor choices for curtains and draperies.

In general, silk can stand heat (it is subjected to very high temperatures when the gum is removed, and most silk dyes are steam set), but does not do well in extreme changes of temperatures, or in overheating through excessive drying.

Some silk yardage and clothing can be hand washed if done carefully. For best results use a mild detergent (like Woolite, or even shampoo) and lukewarm water, then roll the fabric in a towel to absorb the water.

Do not twist — just as you wouldn't twist or pull your hair. Iron dry on a low setting. If you're unsure about washing, check with the manufacturer when possible. Many manufacturers will tell you to dry-clean because it is simpler and yields better results. Wherever possible, you may want to check a swatch first.

Structured silk garments and fragile fabrics should be dry-cleaned to prevent damage. Multi-colour prints or hand-dyed scarves may need to be dry-cleaned to prevent running. You may wish to dry-clean your garment the first time. The steaming process used at the dry-cleaners may also help to further set the dyes.

Moths will attack silk, as well as wool. Store your silk clothing appropriately. As with all fine fabrics, if you plan to store for a long time, you will do best to store in a cotton pillowcase or otherwise surround the silk with a fabric that can breathe. Avoid storing in plastic since this can trap moisture, which can lead to yellowing or the accumulation of mildew. Never use chlorine bleach on silk. It will yellow the fibre and may cause it to breakdown more quickly.

The colours in your silk will undoubtedly fade over time, even when permanent dyes have been used and they have been professionally set. Reds are particularly sensitive to running and fading. Store your silk away from exposure to light, especially direct sunlight. Washing silk may also cause excess dye to discharge. When in doubt, dry-clean the garment or item.

THE ORIGIN OF SILK

Fabric Silk production, or 'Sericulture' as it is known, has a long history. The history of silk is shrouded in mystery. Some historians say that silk originated in China about 10,000 years ago. In China, however, legend has that Empress Si Ling Chi of China was the first one to discover silk more than 5,000 years ago. The story goes that while sitting under a mulberry tree in her palace garden, the Empress was having tea. A cocoon fell into her hot cup of tea from the branches of the mulberry tree. And as she watched, a strong white thread unraveled itself. That accidental process of unravelling the silkworm's cocoon hasn't changed much in the last thousands of years. Cocoons are still dipped in hot water to loosen the tightly woven filaments that make up a silkworm's cocoon. These filaments, or fine threads, are unwound onto a spool. Each cocoon is made up of a filament between 600 and 900 meters long! You could take just one tiny cocoon and unravel its filament, and you could wind it around your house more than eight times. Of course, these filaments are very very thin. Between five and eight of these super-fine filaments are twisted together to make one silk thread. By the 14th century BC, the production of silk gradually developed into an industry in China. In fact it became one of the principal elements of the Chinese economy. Silk was being used for musical instruments, fishing-lines, bowstrings, and even rag paper, the world's first luxury paper.

For more than two thousand years, the Chinese kept the secret of silk to themselves. It was one of the most zealously guarded secrets in history. Anyone found guilty of smuggling silkworm eggs, cocoons, or mulberry seeds was put to death. Silk became the cloth of emperors and royalty and a great source of wealth.

Within the palace, the emperor used to adorn a robe of white silk while outside, he, his principal wife, and the heir to the throne wore yellow, the colour of the earth. Common people were prohibited from wearing silk. The Chinese used silk in many ways. Civil servants were paid in quantities of silk, and it was a way for the rulers to reward their subjects. Silk was also used as currency in domestic as well as foreign trade. After all, barter trade was the most common way of trading. So Chinese traders exchanged their silk in return for spices and jewels bought in India.

Gradually, people from different sections of society began wearing tunics of silk, and silk came into more general use.

THE SECRET OF SILK GETS

A large number of people migrated from China and settled in Korea. They carried with them the secret of silk production. From then on, the secret of silk slowly traveled throughout Asia. It took about five hundred years for the knowledge of silk production to reach India. But once it did, India soon started trading silk with the Persians. However, in less than a hundred years, the Persians too had mastered the art of silk weaving, and started developing their own rich patterns and techniques.

Eventually, silk production became widespread in Europe, but the widespread production of silk hasn't dented China's image as the Silk Giant. Even today, 5,000 years after the first cocoon fell into the Empress' teacup, China is still the world's largest producer of silk.

Gilling

After the carding process a "web" of fibres is formed. The fibres are mostly parallel, however, the sliver has to undergo a further series of straightening processes called "gilling" until the fibres are parallel enough to be passed through the comb. Hooks that are formed in carding are straightened to ensure an effective combing process. The gilling process is performed at two stages: preparation for combing; and finishing after combing.

Fig. Silk Gilling Machine

The purpose of the Preparer gills is to reduce the number of hooks and thus the amount of fibre breakage in the comb which ultimately reduces the noil content.

The purpose of the Finisher gills is to "finish" the final product by blending and evenly distributing the fibres.

Preparing (Gilling)

Gilling, prior to combing (preparing) is a very important step in the process chain. There are a number of purposes for preparation

- Blending of fibres through the doubling process
- Aligning the fibres through the gill head
- Ensuring an even weight per unit length of sliver to facilitate comb setting
- Contentiously, to ensure the hooks produced in carding are aligned the correct way to minimise breakage in the comb. The hooks need to be in the trailing position.

There are several types of gill box, and many variations within types, depending on where in the system drafting occurs.

1 Gilling Principles

However, very simply, a gill box has two sets of rollers and the slivers are dragged from one to another through a bed of moving pinned combs.

The feed rollers pick up the slivers and the delivery rollers, that rotate far faster, draw out the combined slivers into a lighter weight single sliver.

The comb pins are pushed into the sliver and move forward at an even speed with the fibres which controls the acceleration of the fibres between the sets of rollers. When the pins are first thrust into the sliver the feed rollers are still gripping the fibres. The pins are drawn through the fibres and comb them.

At the other end, the delivery rollers pull the fibres faster than the pins are moving so the sliver is drawn through the pins, giving another combing effect. A single sliver comes out of the gill box and is coiled into a can. A variant on the delivery may have two slivers being delivered into a single can. Additionally extra drafting may be achieved by installing a roller drafting system before the gill head. This is often used when the wool has been backwashed or superwashed as it minimises breakage of fibre in the gill head.

The speed differential between the first and second rollers is the draft. Maximum input roller speeds on modern gill boxes is 90m/min whilst output speeds can be as high as 450 m/min. It is normal for a wide range of wool types to have 3 gill passages between the cards and the combs.

Although there is a great deal of science in the measurement of 'hooks" and their effect on top length and romaine, some processors do not consider it important. A hook or bend in the fibre is formed by the action of the metallic card wire on the fibre.

The friction of the fibre being drawn past the wire creates a bend (hook) at an end of the fibre. If this fibre is presented to the comb with its hook in the advanced position the fibre is likely to break.

If the fibre is presented in the trailing position it is likely to be straightened. The difference could be 3mm in length and 0.5% romaine.

The number of gill passages has an impact on the presentation of hooks to the comb. An uneven number presents the majority of hooks in the trailing position whilst an even number would have the opposite effect. In coarse wools > 26ì it maybe immaterial for length and romaine considerations so often a gilling passage is left out. For fine long tops however the amount of gilling and the total draft through the gilling process has a significant impact on the combing results.

The total draft through the preparing area is calculated by multiplying successive drafts. Thus a draft in the first position of 5:1 followed by a draft of 6:1 followed by a draft of 8:1 in the third passage would give a total draft of 240:1. Experimentation has shown that for fine long tops the higher the draft the better the length and the lower the romaine. For very fine wool, a fourth draft has been shown to be beneficial. It is felt that the additional benefits from the parallelisation of fibres overcome the disadvantages of leading hooks after the fourth drafting. Most mills however would not find the fourth draft acceptable in terms of machinery and operator costs.

The gauge on a gill box measures the distance between the drawing off rollers and the faller bed at its closest approach. Between the pins and the drawing off roller there is a space where the fibres are no longer constrained and can move relative to the direction of the acceleration. Normally this does not present a problem as the fibres are long enough either to be influenced by the faller bed or controlled by the drawing off roller. With long fibres the fibre to fibre friction will keep the shorter fibres basically aligned. However when producing short blends i.e. open top <50mm this distance becomes critical. The bulk of the fibres at some point will not be controlled by anything but the fibre to fibre friction and this can lead to large problems. Over drafting of short fibre blends can result in very uneven slivers.

The gilling stage also allows extra processing oil and moisture to be added prior to combing. This is normally applied as an emulsion sprayed between layers of slivers at the first and/or second gill passage.

On high speed gilling machines regain can be increased by approximately 2% on each application. Because of the speed of modern processing machines, regain loss can be as high as 2% on each passage, so moisture needs to be added when possible particularly in dry climate areas.

BACKWASHING

Wool scouring removes most of the contaminants contained in greasy wool but there is always a residual level of wool wax, suint and dirt.

Contaminants may also be redeposited on sliver during processing through a combing mill. This is caused by build up on machines from previous batches.

A spinner may require an exceptionally clean sliver so the sliver may be backwashed. This can be carried out either after preparing or after the final finishing gill.

There are purpose built machines that feed the sliver through a weak detergent water solution then through a drier or a Superwash line can be employed to do the job using only water and detergent.

Backwashing adds large processing costs to a batch so is only carried out when absolutely necessary.

SUPERWASHING

Scales on wool fibres entangle and cause products to felt when machine washed so a treatment was developed to prevent this happening.

Finished wool sliver is treated with chlorine and then coated with a resin. The chlorine removes a small portion of the offending wool scales and the resin coating forms a smooth fibre that will no longer felt.

This is a permanent process that will last the life of a product.

8

Wool Fibres

Wool, common name applied to the soft, curly fibres obtained chiefly from the fleece of domesticated sheep, and used extensively in textile manufacturing. Wool may be differentiated from hair mainly by the nature of the scales that cover the outer surface of each fibre. Wool scales are numerous, minute, and pointed and are attached only at their bases; thus the fibres interlock under pressure. The number of scales varies with the fineness and curliness of the fibre. Because of its crimp, or curl, wool has considerable resilience. This quality, together with its high tensile strength and elasticity, gives fine woolen fabrics the ability to retain shape better than cloth made from other natural fibres. Other characteristics of wool, which make it especially desirable for clothing, are its lightness, its ability to absorb moisture, and its insulating properties.

SOURCES OF WOOL

Wild species of sheep have a short, woolly undercoat covered by long, coarse, straight hair. The hair has entirely disappeared in domestic breeds; selective breeding has improved both the quality and abundance of wool. Wool production of sheep is also influenced by nutrition, climate, and care. Neglected domesticated animals have a tendency to produce hair instead of wool.

The fleece of sheep raised for wool is generally shorn once yearly, in the spring or early summer. In regions where the climate is warm throughout the year, shearing may occur twice annually.

The fleece is cut close to the skin, usually with mechanical shears, and removed in one piece. The average weight of the fleece shorn from better wool-producing breeds is about 4.5 kg (about 10 lb). Small amounts of wool come from the skins of lambs slaughtered for meat.

Wool from different parts of the same fleece varies greatly in length of fibre, fineness, and structure. Wool from the shoulders and sides of the sheep is usually superior to that from other parts of the body. Quality varies also with the different breeds of sheep.

Pure merino sheep bear the finest wool, and this type has been crossed with hardy sheep producing longer, coarser wool. About 40 percent of the world's total wool production consists of merino wool, and about 43 percent is obtained from crossbred varieties.

The remainder is composed largely of carpet wools, which are derived mostly from special types of sheep and are employed in the manufacture of blankets, carpets, and tapestries. In addition, a small quantity of fine wool used for apparel is obtained from animals other than sheep. These wool-bearing animals include the camel, the alpaca, the Angora and Kashmîr goats, the llama, and the vicuña.

Types and Labeling of Wool

The value of wool on the market depends primarily on fineness and length of fibre. Strength, elasticity, amount of crimp, and uniformity are also considered. Two different systems are followed in wool processing, the woolen system and the worsted system. In the woolen system, the fibres are carded and then spun. In the worsted system, the fibres proceed to a combing process, which separates the long from the short fibres.

The long fibres are then formed into smooth compact strands ready for spinning into yarn. Worsteds are not napped but are finished with a smooth surface. Uniform lengths of relatively fine fibres are very important in the worsted system, because short fibres are difficult to spin. In the woolen system, on the other hand, fibres of mixed lengths may be used. Fine wools are classified according to the length of fibre. Longer fibres used in the manufacture of worsteds are called combing wool; shorter fibres employed for yarns used in soft, woven fabrics are known as carding wool.

In 1939, to protect the consumer, the U.S. Congress passed the Wool Products Labeling Act. This law requires that all fabric containing wool carry identification indicating the percentage of wool in the cloth. The act also specifies the proper legal description for the fibre according to category, such as new, reprocessed, or reused wool. New wool, or virgin wool, is wool that never before has been used in manufacture.

Reprocessed wool refers to fibre reclaimed from woven scraps and clips accumulated from the cutting table, from samples, and from mill ends.

These remnants, never previously used, are reconverted into a fibrous state and then reprocessed into the manufactured product. Reused wool is wool fibre recovered from products used by consumers, respun, and rewoven. This category is less important now than formerly, because of competition from synthetics.

PROCESSING OF WOOL

SHEARING

After shearing, the wool is separated into four main categories: fleece (which makes up the vast bulk), broken, bellies, and locks. The quality of fleeces is determined by a technique known as wool classing, whereby a qualified person called a wool classer groups wools of similar gradings together to maximize the return for the farmer or sheep owner. In Australia and New Zealand, before being auctioned, all Merino fleece wool is objectively measured for micron, yield (including the amount of vegetable matter), staple length, staple strength, and sometimes color and comfort factor.

Scouring

Wool straight off a sheep, known as "greasy wool" or "wool in the grease", contains a high level of valuable lanolin, as well as dirt, dead skin, sweat residue, pesticide, and vegetable matter.

Before the wool can be used for commercial purposes, it must be scoured, a process of cleaning the greasy wool. Scouring may be as simple as a bath in warm water, or as complicated as an industrial process using detergent and alkali, and specialized equipment.

In commercial wool, vegetable matter is often removed by chemical carbonization. In less-processed wools, vegetable matter may be removed by hand, and some of the lanolin left intact through use of gentler detergents. This semigrease wool can be worked into yarn and knitted into particularly water-resistant mittens or sweaters, such as those of the Aran Island fishermen. Lanolin removed from wool is widely used in cosmetic products such as hand creams.

Quality

The quality of wool is determined by its fiber diameter, crimp, yield, colour, and staple strength. Fiber diameter is the single most important wool characteristic determining quality and price.

Merino wool is typically 3–5 inches in length and is very fine (between 12 and 24 microns). The finest and most valuable wool comes from Merinohoggets. Wool taken from sheep produced for meat is typically more coarse, and has fibers 1.5 to 6 inches (38 to 150 mm) in length. Damage or breaks in the wool can occur if the sheep is stressed while it is growing its fleece, resulting in a thin spot where the fleece is likely to break.

Wool is also separated into grades based on the measurement of the wool's diameter in microns and also its style. These grades may vary depending on the breed or purpose of the wool. For example:

- <15.5 - Ultrafine Merino
- 15.6–18.5 – Superfine Merino
- 18.6–20 – Fine Merino
- 20.1–23 – Medium Merino
- 23< - Strong Merino
- Comeback: 21–26 microns, white, 90–180 mm long

- Fine crossbred: 27–31 microns, Corriedales, etc.
- Medium crossbred: 32–35 microns
- Downs: 23–34 microns, typically lacks luster and brightness. Examples, Aussiedown, Dorset Horn, Suffolk, etc.
- Coarse crossbred: 36> microns
- Carpet wools: 35–45 microns.

Any wool finer than 25 microns can be used for garments, while coarser grades are used for outerwear or rugs. The finer the wool, the softer it is, while coarser grades are more durable and less prone to pilling.

The finest Australian and New Zealand Merino wools are known as 1PP, which is the industry benchmark of excellence for Merino wool 16.9 microns and finer. This style represents the top level of fineness, character, colour, and style as determined on the basis of a series of parameters in accordance with the original dictates of British Wool as applied today by the Australian Wool Exchange (AWEX) Council. Only a few dozen of the millions of bales auctioned every year can be classified and marked 1PP.

History

Wild sheep were more hairy than woolly. Although sheep were domesticated nine to eleven thousand years ago, archaeological evidence from statuaryfound at sites in Iran suggests that selection for woolly sheep may have begun around 6000 BC, with the earliest woven wool garments having only been dated to two to three thousand years later. Woolly-sheep were introduced into Europe from the Near East in the early part of the 4th millennium BC. The oldest known European wool textile, ca. 1500 BC, was preserved in a Danish bog.

Prior to invention of shears – probably in the Iron Age – the wool was plucked out by hand or by bronze combs. In Roman times, wool, linen, and leatherclothed the European population; the cotton of India was a curiosity that only naturalists had heard of; and silk, imported along the Silk Road from China, was an extravagant luxury. Pliny the Elder records in his Natural History that the reputation for producing the finest wool was enjoyed byTarentum, where selective breeding had produced sheep with a superior fleece, but which required special care.

In medieval times, as trade connections expanded, the Champagne fairs revolved around the production of wool cloth in small centers such as Provins; the network that the sequence of annual fairs developed meant that the woollens of Provins might find their way to Naples, Sicily, Cyprus, Majorca, Spain, and even Constantinople. The wool trade developed into serious business, the generator of capital. In the 13th century, the wool trade was the economic engine of the Low Countries and central Italy; by the end of the following century, Italy predominated, though in the 16th century Italian production turned to silk. Both pre-industries were based on English raw wool exports – rivaled only by the sheepwalks of Castile, developed from the 15th century – which were a significant source of income to the English crown, which from 1275 imposed an export tax on wool called the "Great Custom". The importance of wool to the English economy can be shown by the fact that since the 14th century, the presiding officer of the House of Lords has sat on the "Woolsack", a chair stuffed with wool.

Economies of scale were instituted in the Cistercian houses, which had accumulated great tracts of land during the 12th and early 13th centuries, when land prices were low and labour still scarce. Raw wool was baled and shipped from North Sea ports to the textile cities of Flanders, notably Ypres andGhent, where it was dyed and worked up as cloth. At the time of the Black Death, English textile industries accounted for about 10% of English wool production; the English textile trade grew during the 15th century, to the point where export of wool was discouraged. Over the centuries, various British laws controlled the wool trade or required the use of wool even in burials. The smuggling of wool out of the country, known as owling, was at one time punishable by the cutting off of a hand. After the Restoration, fine English woollens began to compete with silks in the international market, partly aided by the Navigation Acts; in 1699, the English crown forbade its American colonies to trade wool with anyone but England herself.

A great deal of the value of woollen textiles was in the dyeing and finishing of the woven product. In each of the centers of the textile trade, the manufacturing process came to be subdivided into a collection of trades, overseen by an entrepreneur in a system called by the English the "putting-out" system, or "cottage industry", and the *Verlagssystem* by the Germans. In this system of producing wool cloth, until recently perpetuated in the production of Harris tweeds, the entrepreneur provides the raw materials and an advance, the remainder being paid upon delivery of the product. Written contracts bound the artisans to specified terms. Fernand Braudel traces the appearance of the system in the 13th-century economic boom, quoting a document of 1275. The system effectively bypassed the guilds' restrictions.

Before the flowering of the Renaissance, the Medici and other great banking houses of Florence had built their wealth and banking system on their textile industry based on wool, overseen by theArte della Lana, the wool guild: wool textile interests guided Florentine policies. Francesco Datini, the "merchant of Prato", established in 1383 an *Arte della Lana* for that small Tuscan city. The sheepwalks of Castile shaped the landscape and the fortunes of the *meseta* that lies in the heart of the Iberian peninsula; in the 16th century, a unified Spain allowed export of Merino lambs only with royal permission. The German wool market – based on sheep of Spanish origin – did not overtake British wool until comparatively late. The Industrial Revolution introduced mass production technology into wool and wool cloth manufacturing. Australia's colonial economy was based on sheep raising, and the Australian wool trade eventually overtook that of the Germans by 1845, furnishing wool for Bradford, which developed as the heart of industrialized woollens production. Due to decreasing demand with increased use of synthetic fibers, wool production is much less than what it was in the past.

The collapse in the price of wool began in late 1966 with a 40% drop; with occasional interruptions, the price has tended down. The result has been sharply reduced production and movement of resources into production of other commodities, in the case of sheep growers, to production of meat.

Superwash wool (or washable wool) technology first appeared in the early 1970s to produce wool that has been specially treated so it is machine washable and may be tumble-dried. This wool is produced using an acid bath that removes the "scales" from the fiber, or by coating the fiber with a polymer that prevents the scales from attaching to each other and causing shrinkage. This process results in a fiber that holds longevity and durability over synthetic materials, while retaining its shape.

In December 2004, a bale of the world's finest wool, averaging 11.8 micron, sold for $3,000 per kilogram at auction in Melbourne, Victoria. This fleece wool tested with an average yield of 74.5%, 68 mm long, and had 40 newtons per kilotex strength. The result was A$279,000 for the bale. The finest bale of wool ever auctioned was sold for a seasonal record of 269,000 cents per kilo during June 2008. This bale was produced by the Hillcreston Pinehill Partnership and measured 11.6 microns, 72.1% yield and had a 43 Newtons per kilotex strength measurement. The bale realized $247,480 and was exported to India.

During 2007, a new wool suit was developed and sold in Japan that can be washed in the shower, and dries off ready to wear within hours with no ironing required. The suit was developed using Australian Merino wool, and it enables woven products made from wool, such as suits, trousers and skirts, to be cleaned using a domestic shower at home.

In December 2006, the General Assembly of the United Nations proclaimed 2009 to be the International Year of Natural Fibres, so to raise the profile of wool and other natural fibers.

Production

Global wool production is approximately 1.3 million tonnes per year, of which 60% goes into apparel. Australia is the leading producer of wool which is mostly from Merino sheep. New Zealand is the second-largest producer of wool, and the largest producer of crossbred wool. China is the third-largest producer of wool. Breeds such as Lincoln, Romney, Tukidale, Drysdale and Elliotdale produce coarser fibers, and wool from these sheep is usually used for making carpets.

MARKETING

Australia

About 85% of wool sold in Australia is sold by open cry auction. 'Sale by sample' is a method in which a mechanical claw takes a sample from each bale in a line or lot of wool. These grab samples are bulked, objectively measured, and a sample of not less than 4 kg is displayed in a box for the buyer to examine. The Australian Wool Exchange conducts sales primarily in Sydney, Melbourne, Newcastle, and Fremantle. There are about 80 brokers and agents throughout Australia.

About 7% of Australian wool is sold by private treaty on farms or to local wool-handling facilities. This option gives wool growers benefit from reduced transport, warehousing, and selling costs. This method is preferred for small lots or mixed butts in order to make savings on reclassing and testing.

About 5% of Australian wool is sold over the internet on an electronic offer board. This option gives wool growers the ability to set firm price targets, reoffer passed-in wool and offer lots to the market quickly and efficiently. This method works well for tested lots, as buyers use these results to make a purchase. About 97% of wool is sold without sample inspection; however, as of December 2009, 59% of wool listed had been passed in from auction. Growers through certain brokers can allocate their wool to a sale and at what price their wool will be reserved.

Sale by tender can achieve considerable cost savings on wool clips large enough to make it worthwhile for potential buyers to submit tenders. Some marketing firms sell wool on a consignment basis, obtaining a fixed percentage as commission.

Forward selling: Some buyers offer a secure price for forward delivery of wool based on estimated measurements or the results of previous clips. Prices are quoted at current market rates and are locked in for the season. Premiums and discounts are added to cover variations in micron, yield, tensile strength, etc., which are confirmed by actual test results when available.

Another method of selling wool includes sales direct to wool mills.

Other Countries

The British Wool Marketing Board operates a central marketing system for UK fleece wool with the aim of achieving the best possible net returns for farmers.

Less than half of New Zealand's wool is sold at auction, while around 45% of farmers sell wool directly to private buyers and end-users.

United States sheep producers market wool with private or cooperative wool warehouses, but wool pools are common in many states. In some cases, wool is pooled in a local market area, but sold through a wool warehouse. Wool offered with objective measurement test results is preferred. Imported apparel wool and carpet wool goes directly to central markets, where it is handled by the large merchants and manufacturers.

Yarn

Virgin wool is wool spun for the first time. Shoddy or recycled wool is made by cutting or tearing apart existing wool fabric and respinning the resulting fibers. As this process makes the wool fibers shorter, the remanufactured fabric is inferior to the original. The recycled wool may be mixed with raw wool, wool noil, or another fiber such as cotton to increase the average fiber length. Such yarns are typically used as weft yarns with a cotton warp. This process was invented in the Heavy Woollen District of West Yorkshire and created a micro-economy in this area for many years.

Ragg is a sturdy wool fiber made into yarn and used in many rugged applications such as gloves. Worsted is a strong, long-staple, combed wool yarn with a hard surface. Woollen is a soft, short-staple, carded wool yarn typically used for knitting. In traditional weaving, woollen weft yarn (for softness and warmth) is frequently combined with a worsted warp yarn for strength on the loom.

Uses

In addition to clothing, wool has been used for blankets, horse rugs, saddle cloths, carpeting, felt, wool insulation (also see links) and upholstery. Wool felt covers piano hammers, and it is used to absorb odors and noise in heavy machinery and stereo speakers. Ancient Greeks lined their helmets with felt, and Roman legionnaires used breastplates made of wool felt. Wool has also been traditionally used to cover cloth diapers. Wool fiber exteriors are hydrophobic (repel water) and the interior of the wool fiber ishygroscopic (attracts water); this makes a wool garment able to cover a wet diaper while inhibiting wicking, so outer garments remain dry. Wool fluted and treated with lanolin is water resistant, air permeable, and slightly antibacterial, so it resists the buildup of odor. Some modern cloth diapers use fluted wool fabric for covers, and there are several modern commercial knitting patterns for wool diaper covers. Initial studies of woollen underwear have found it prevented heat and sweat rashes because it more readily absorbs the moisture than other fibers. Merino wool has been used in baby sleep products such as swaddle baby wrap blankets and infant sleeping bags. Wool is an animal protein, and as such it can be used as a soil fertilizer, being a slow-release source of nitrogen and ready-made amino acids. Researchers at the Royal Melbourne Institute of Technology school of fashion and textiles have discovered a blend of wool and kevlar, the synthetic fiber widely used in body armor, was lighter, cheaper and worked better in damp conditions than kevlar alone. Kevlar, when used alone, loses about 20% of its effectiveness when wet and therefore required an expensive waterproofing process. Wool increased friction in a vest with 28–30 layers of fabric, to provide the same level of bullet resistance as 36 layers of Kevlar alone.

Events

A buyer of Merino wool, Ermenegildo Zegna, has offered awards for Australian wool producers. In 1963, the first Ermenegildo Zegna Perpetual Trophy was presented in Tasmania for growers of "Superfine skirted Merino fleece". In 1980, a national award, the Ermenegildo Zegna Trophy for Extrafine Wool Production, was launched. In 2004, this award became known as the Ermenegildo Zegna Unprotected Wool Trophy. In 1998, an Ermenegildo Zegna Protected Wool Trophy was launched for fleece from sheep coated for around nine months of the year.

In 2002, the Ermenegildo Zegna Vellus Aureum Trophy was launched for wool that is 13.9 micron and finer. Wool from Australia, New Zealand,Argentina, and South Africa may enter, and a winner is named from each country. In April 2008, New Zealand won the Ermenegildo Zegna Vellus Aureum Trophy for the first time with a fleece that measured 10.8 microns. This contest awards the winning fleece weight with the same weight in gold as a prize, hence the name.

In 2010 an ultra-fine, 10 micron fleece, from Windradeen, near Pyramul, New South Wales set a new world record in the fineness of wool fleeces when it won the Ermenegildo Zegna Vellus Aureum International Trophy. Since 2000, Loro Piana has awarded a cup for the world's finest bale of wool that produces just enough fabric for 50 tailor-made suits. The prize is awarded to an Australian or New Zealand wool grower who produces the year's finest bale.

The New England Merino Field days which display local studs, wool, and sheep are held during January, every two years (in even numbered years) around the Walcha, New South Wales district. The Annual Wool Fashion Awards, which showcase the use of Merino wool by fashion designers, are hosted by the city of Armidale, New South Wales in March each year. This event encourages young and established fashion designers to display their talents. During each May, Armidale hosts the annual New England Wool Expo to display wool fashions, handicrafts, demonstrations, shearing competitions, yard dog trials, and more.

In July, the annual Australian Sheep and Wool Show is held in Bendigo, Victoria. This is the largest sheep and wool show in the world, with goats and alpacas as well as woolcraft competitions and displays, fleece competitions, sheepdog trials, shearing, and wool handling. The largest competition in the world for objectively-measured fleeces is the Australian Fleece Competition, which is held annually at Bendigo. In 2008, there were 475 entries from all states of Australia with first and second prizes going to the Northern Tablelands, New South Wales fleeces.

WOOL PRODUCTION

Australia is the world's largest producer of raw wool, growing about 30 percent of the total world supply. Other leading producers are China, New Zealand, Iran, the United Kingdom, Argentina, South Africa, and Uruguay. In 2005 annual U.S. wool production was 16,885 metric tons when weighed before cleaning; most of the wool came from sheep raised in Texas, Wyoming, California, Colourado, Montana, and South Dakota. All the carpet wool and more than one-half of the apparel wool consumed in U.S. mills is imported from other countries.

THE PRODUCTION OF WOOL

The United States produces annually something over 300,000,000 pounds of wool as it comes from the sheep's back. The people each year use, on an average, between five and six pounds for every man, woman, and child; hence there must be imported almost as much as we produce at home. The production within the United States in 1910 was 321,400,000 pounds, while the importations amounted to 252,000,000 pounds. The sources of our raw wool will prove more significant in the future, for according to the census of 1910, there were then about 14 per cent fewer sheep in this country than there were in 1900, ten years before. During these ten years the number of sheep fell off in more than three-fourths of the states in the Union. In some states the drop in numbers was as much as half. Evidently if we allow this decrease in production to continue, we shall have to rely more and more upon foreign countries for our supply.

Why wool production is falling off in the United States. It is well to know something about the conditions of sheep and wool production in this country so that this tendency to discontinue sheep raising may be traced intelligently. Possibly some of the causes for this tendency may be removed or at least modified. It may be that some of them will become inoperative in the future. In any case the student should intelligently follow this movement during the coming years.

Not profitable.-The fundamental reason for the lessening of sheep in America is that the farmers find that under the present methods of farming they can make more money producing something else. The average market value of a sheep in this country in 1910 was about $4.40. The cost of keeping sheep averages about $4.00 per head per year, counting all expenses, value of investment, etc. The principal products are wool and lambs. The average weight of the fleece of an American sheep is less than seven pounds. At an average sales price of twenty-five cents a pound, the wool brings about $1.75. Flocks that double in number in a year are considered very good. If this increase in lambs were sold at the average value of lambs as given in the census for 1910, then there would be a product of about $2.30 for each lamb. Each sheep in the flock could then be credited with a fleece at $1.75 and a lamb at $2.30, or a total of $4.05. This leaves a net profit of five cents a sheep-too narrow a margin in view of the facts that only the better grades of wool sell for twenty-five cents a pound, that the price fluctuates considerably, and that the cost of keeping sheep varies from year to year. It is easy to see why many farmers lose money on sheep raising. The margin of profit is never very wide. Careful accounts of nine typical Minnesota flocks kept by the Minnesota Agricultural College showed for each sheep an average cost of $4.13 and an average income of $4.38.

Western ranges raise sheep at considerably less cost, but the product on each sheep is generally less also. The figures that have been given should be considered only as averages. Few actual cases will be found exactly the same as these. The point emphasized is the narrow profit margin in sheep production in this country. Many who have scientifically studied sheep raising say that it is possible to raise sheep in this country at a satisfactory profit. The fact is that while certain farmers are able to raise sheep profitably, their methods and their knowledge of the business are not general. It is one of the chief purposes of agricultural colleges the country over to teach better methods of production; some of these colleges have made special study of this particular subject of sheep and wool production.

Depredations of dogs.-Not only the financial consideration prevents farmers from raising sheep. Strange as it may seem, sheep raising has fallen off considerably in several portions of this country because of the depredations of dogs. Vermont is an example, according to some who have grown sheep there. The same is true of Massachusetts, New York, and even Wisconsin. In Kansas, where there are more dogs than sheep, the sheep owners declare that the owners of dogs cannot be made to believe that their dogs are the culprits; hence trouble arises. Where dog owners outnumber the raisers of sheep, the latter frequently stop raising sheep. It should be added that much of the sheep destruction by dogs is due to stray curs, an evil that a good licensing system would remedy.

Competition of the dairy industry.-Another factor checking the production of sheep is the growth of the dairy industry. In southern Wisconsin, formerly a great sheep section, dairy cattle have been substituted to the full limit of pasturage and feed. The Wisconsin College of Agriculture maintains that this is a mistake. They assert that if the farmer would raise fewer cows and would add small flocks of sheep, the net profits would be as great and the farm kept in better condition.

High cost of fencing.-Another drawback to sheep raising in the northern part of Wisconsin is the high first expense for fencing. Cattle can be kept within a pasture by the use of only three barbed wires placed at the proper height; whereas sheep require five wires, or better still, woven wire fencing. The immediate expense of such fencing deters new farmers from going into sheep raising. The narrow margin of profit, the dangers of sheep diseases, and the lack of knowledge about sheep are some of the other reasons why sheep raising is on the decline in the Middle West.

Methods of wool production in this country.-The methods of wool production differ widely in the different parts of the country. For example, in the West, sheep raising is conducted as a distinct occupation on large ranches, huge flocks of sheep being raised on a single ranch containing its thousands of acres. In the eastern part of the country the industry of sheep raising is carried an rather incidentally on grain or dairy farms. In Wisconsin the average number of sheep on the farms that raise sheep at all is less than twenty. This may be taken as typical of sheep production in all parts of the country except in the sheep-ranching area. The sheep ranches require special labour, sheep herders, camp tenders, and shearers, an average of one man for every 1,500 head of sheep, and extra help during lambing and shearing times. In the dairy and grain-farming states, the sheep are usually given only a part of the time of the farmer, allowed to run in the pastures with the cattle, and stabled just as the cattle are in the winter time. In the West, the sheep ranch land is used for nothing else. In fact close grazing unfits the land for other livestock pasturage for several seasons. Sheep ranching has frequently paid well in the past, but is, nevertheless, a hazardous and lonesome occupation. The sheep herder stays out for long periods of time with his flocks, seeing no human being save the camp tender. Few men can stand the strain without undue mental suffering. Most of the herders are Mexicans and Indians, who seem less susceptible to this danger.

Possibilities of loss an sheep ranches.-There is always possibility of great loss of sheep from storms during the winter. During blizzards whole flocks of many thousands of sheep may be killed. Poisonous plants kill some sheep every year. Coyotes and wildcats, especially the former, constantly follow the ranch flocks, stealing in and killing sheep whenever the herder with his rifle is off his guard. In times of excessive drought, both the feed and water may disappear; whereupon the sheep must be driven long distances to new pastures and to water. Many sheep are likely to fall dead by the wayside on these journeys. There is usually considerable loss of both ewes and young lambs at lambing time. For weeks after the lambs are born, there is constant likelihood of the lambs being separated from the mothers, and starving to death. At least 10 per cent of the lambs die thus on the sheep ranches. Large numbers of sheep are often shipped eastward and divided up among eastern farmers for the purpose of fattening them for market, with probable losses at every move. Sheep ranching has almost as many chances as gambling. During the last few years, therefore, there has been a steady decrease in the number and size of these ranches. Irrigation and dry farming have crept in. The sheep rancher has been compelled either to retreat to more arid regions, or to sell out his flocks, very frequently the latter. We may not look forward to any increase in sheep ranching in this country. Sheep and wool production will grow only with the spread of diversified farming, and in the humid regions.

Shearing.-Shearing is now usually done in May. June shearing, formerly common, is now thought less practicable, the sheep doing better and the wool product being finer from the earlier shearing. The average fleece in Wisconsin weighs from seven to eight and one-half pounds, a fair average for the varieties raised, mainly Shropshires and Oxfords.

The shearing is done either by hand shears or by machine clippers. Since the introduction of the machine wool clippers, the sheep with folds in their skin have not been popular. The older merinos had deep folds, and sheep raisers everywhere tried to breed so as to make them still deeper; they reasoned that the deeper the folds the greater the skin surface, and the greater the skin surface, the greater the amount of wool on each sheep. Now farmers everywhere desire the sheep with smooth skin, preferring rather to increase the size of the body by crossbreeding the merinos and other fine wool sheep with the large, bony, long-wool types.

The fleece.-If the shearing is done well, the wool from the sheep holds together in one large sheet. This sheet, called the fleece, is rolled up and tied, with the dirty outer side of the wool turned in. Sometimes two fleeces are tied together in the same bundle. The fleeces are either stored by the farmers in the hope that prices may rise, or else immediately hauled by wagon loads to market.

Wool washing.-In some localities the wool is washed before being hauled to market, frequently while on the sheep's back, by giving the sheep a bath in a tank, but this practice seems to be on the decrease. The improved washing machinery found in wool manufactories has made it unprofitable for the farmer to wash wool by hand in his ordinary tubs or tanks. The result from wool washing on the farm is so uneven that there can be no assurance of proper remuneration for the work.

THE GEOGRAPHY OF WOOL PRODUCTION

General facts.-The facts regarding the production of raw wool in this country, the number of sheep, and our imports of wool may be best described by means of a report which was prepared by the U. S. Tariff Board in 1911.

Primary wool markets.-The raw wool markets of the world are of two classes, primary wool markets and concentration points. First, there are primary markets where the wool changes hands from the grower to dealer or other buyer. These primary markets are numerous and widely scattered, especially in this country. There are a few places, however, which because of their location in the sheep-range country have become prominently known as primary markets. Among these are Billings, Big Timber, and Great Falls in Montana, and San Francisco, San Diego, and Portland on the Pacific Coast. Concentration points.-More important still are the concentration points, where raw wool is handled in large quantities, bought and sold by dealers and manufacturers, exporters and importers. London is the greatest wool market in the world. Here wools from all over the world, particularly the finer grades, are brought and sold to buyers from all parts of England, the Continent, and even the United States. Sydney in New South Wales, Australia, comes second in quantity and value of wool handled.

The growth of Sydney as a wool market has been rapid during the last few years. Formerly it was a point of export for London and other European markets, but now buyers are coming to Sydney from all manufacturing countries, from England, Scotland, Germany, France, Belgium, Austria, Italy, Japan, and the United States. Sydney now markets more merino wool than even London. It is also a primary market for large numbers of sheep in New South Wales. Some of the wool warehouses found there cover six to eight acres and are from five to six floors high. Most of them are equipped with the most modern arrangements for baling, for storage, and for keeping the wool in the proper condition.

The other principal markets of Australia are Melbourne, Geelong, Adelaide, and Brisbane. Cape Town in South Africa, Buenos Aires in Argentina, and Montevideo in Uruguay are other highly important wool markets.

The carpet wools are first marketed in the countries where produced. Among the important points from which America and England get carpet wools are Constantinople, Bagdad, Aleppo, and Smyrna, all in Turkish dominions; Karachi and Bombay in India; Tientsin and Shanghai in China; Urga and Biisk in Mongolia; and Rostoff, Odessa, and Moscow in Russia.

The chief raw wool market in the United States is Boston. Philadelphia comes second, while Chicago and New York are also important.

Centralization of wool manufacturing.-Wool manufacturing is very widely distributed in one form or another. In only the modern, highly improved machine processes is a decided centralization of production found. A few of the more important wool goods producing centers of the world are considered here.

United States.-In worsted goods Massachusetts ranks first among the American states; then come Pennsylvania and Rhode Island. Lawrence, Massachusetts; Providence, Rhode Island; Philadelphia, Pennsylvania; and Cleveland, Ohio, are the noted worsted producing cities of the country. Pennsylvania has the largest number of woollen mills, but Massachusetts produces the greatest amount of woollen cloth.

Woollen goods are more widely produced than are the worsteds. Carpet and rug weaving is practically confined to Pennsylvania, New York, and Massachusetts. Of hosiery and knit goods, New York produces the most; Pennsylvania ranks second; Massachusetts, Illinois, Ohio, and Wisconsin follow in the order given.England.-In England nearly three-fourths of the woollen industries are found in the western part of Yorkshire in or near the city of Bradford. Bradford is important as a distributing centre of raw materials, as the centre of worsted combing and spinning, and as the centre for the production of worsted stuffs or dress goods for women's wear. Huddersfield is the centre for the production of the best worsted goods for men's wear.

Fancy cheap woollens made in imitation of Scotch tweeds are made in the Colne Valley near Huddersfield. Colne Valley woollens are known the world over as cheap materials.

Heavy woollens, overcoatings, cheap cloalcings, blankets, and army cloths are made in and near Dewsbury and Batley. This is the centre of the shoddy and rag business, and much of the reclaimed wool, by skilful blending and manipulation, is turned into cheap but serviceable fabrics. Fabrics made at Morley are the lowest quality made. Leeds, once noted for its broadcloths, has now taken up the manufacture of worsted coatings, serges, etc. Halifax is best known as a producer of worsted yarn. Rochdale in Lancashire is important for its flannels, Stroud in western England has the reputation of making the finest woollens in the world. Kidderminster is the most important carpet centre; while Leicester, parts of Nottingham, and Derby are important in the production of hosiery.

Scotland.-In Scotland the manufacture of Scotch tweeds is an important industry, Galashiels being the most important centre.

Germany.-Wool manufactories are found all over the German Empire. There is no such single important centre as Bradford, England; furthermore the various localities each produce both woollens and worsteds. Some of the greatest manufacturing cities are Blumenthal, Hamburg, Leipzig, Dohren, Mylau, Plauen, Muhlhausen, Gera, Aachen, Breslau, and Forst. Among the nations, Germany is the third greatest producer of woollen goods in the world.

France.-In France, the Department of the North, which borders on Belgium, is the most important wool-manufacturing centre. Nearly all the combing is done here, and most of the spinning and weaving as well. Roubaix is the chief city in the business. In a way its leadership is comparable to that of Bradford, England, but, of course, on a smaller scale. Tourcoing and Fourmies, other towns in the same department, are also important.

The town of Elboeuf in Normandy compares with Leeds in England in the manufacture of worsteds. It does a miscellaneous business in both woollens and worsteds. Vienne on the Rhone River is the centre of shoddy cloth manufacture, and the manufacture of blankets is largely confined to the surrounding towns. Mazamet does a remarkable business in the production of pulled wools from South American and Cape sheep skins. Wool velvets are produced in the Department of Somme and plushes in the Department of Aisne.

WOOL MANUFACTURING

The first step in the processing of wool preparatory to cloth making is the sorting of the fibres. As each fleece is pulled apart, the fibres are sorted into separate piles of similar nature and characteristics, mainly on the basis of fineness, length, and freedom from defects.

After sorting, the fibres are cleaned, because wool contains an oily substance, called yolk, consisting of lanolin and suint, which is dried perspiration. The wool is scoured by a soap-alkali preparation or other detergent or solvent to remove the grease and other foreign substances.

The scoured wool fibres are disentangled and drawn straight into a continuous form in a process called carding, which employs a carding machine.

In this process, the fibres are passed between pairs of rotating cylinders covered with a material called card clothing, which contains fine, pliable wire teeth. The wool emerges from the rollers in the form of a thin film known as the web. The processing of the web varies according to whether woolen or worsted yarn is to be produced.

Fig. Wool Manufacturing Machine

Woolen fabrics, such as tweeds, are woven from bulky yarns containing short wool fibres arranged at random, so that the fabric is relatively thick and has a fuzzy surface. Worsted fabrics, such as gabardines, are woven from yarns composed of longer, thinner wool fibres, tightly twisted for a smooth surface.

For the production of woolen yarn, the web is split into soft, thin strands called roving, and then it is ready to be spun. For worsted yarn, the web is processed by machines that straighten the fibres, make them parallel, and remove all the short pieces. The resultant thick strand is then condensed by passing it through several machines until a very thin strand of worsted roving is obtained. From this type of roving can be spun the smooth yarn that produces firm, smooth-textured worsted fabrics.

SHEEP

Sheep, common name for a collection of grazing mammals that may be either wild or domesticated; the domesticated varieties are the most widely distributed kind of domestic animal, found in nearly all countries. Wild species are found in some areas of the world: the bighorn in western North America; Dall sheep in northern Canada and Alaska; the mouflon in Mediterranean countries; the Asian mouflon in western Asia; the urial Afghanistan and Pakistan; and the argali in eastern Asia.

Sheep were probably domesticated about 11,000 years ago in what is now northern Iraq, and they are thought to have descended primarily from the mouflon, although recent evidence suggests that species such as the Asian mouflon contributed to some of the modern breeds.

Characteristics

Sheep are even-toed, hoofed animals. They are cud-chewing animals with the upper incisor teeth missing and with a four-compartmented stomach.

They have paired, hollow, unbranched horns that are not shed. The horns of the adult male, or ram, are massive and spirally curved. The horns of the adult female, or ewe, are short and only slightly curved.

Sheep typically have a long, fairly narrow muzzle and pointed ears. The length of the head and body averages about 1.5 m (about 5 ft), with a short tail, and an adult may weigh 75 to 200 kg (165 to 440 lb).

In the wild, the animals are nimble runners and climbers. The female bears up to three young after a gestation period of about 150 days. Sheep live as long as 20 years.

Besides providing pelts and wool for clothing and carpets, meat in the form of lamb and mutton, and milk for drinking and cheesemaking, sheep are used to a limited extent as pack animals, and the wild species are hunted as game.

Several distinct types and more than 800 breeds of domesticated sheep have been developed. The breeds are adapted to environments that range from desert to tropical conditions.

Fine-Wool Breeds

Sheep bred for their fine wool account for nearly half the world sheep population. They are adapted to semiarid conditions and are characterized as medium in size, with the ability to produce large amounts of wool fibres 20 micrometers or less in diameter.

Found extensively in Australia, New Zealand, South America, and the western United States, most sheep of this type belong to the merino breed, which originated in the Mediterranean area and became concentrated in Spain. The breed, however, has been modified and adapted to the conditions prevalent in various countries, and the different subtypes are usually called merinos combined with the name of their adopted country.

MUTTON-TYPE SHEEP

Mutton-type sheep, which consist of medium- and long-wool breeds, have been bred primarily for their meat. They account for about 15 percent of the world sheep population. Among the medium-wool breeds are the Suffolk, Hampshire, Shropshire, Southdown, Dorset, Isle-de-France, Cheviot, and Oxford; and long-wool breeds, the largest of all breeds, include the Leicester, Lincoln, Cotswold, and Romney Marsh. In the United States, long-wool breeds are used extensively as sires with fine-wool and crossbred females for the production of market lambs.

Many attempts have been made to obtain the superior wool characteristics of fine-wool breeds with the meat-producing abilities of other breeds. These have resulted, in recent times, in the development of several new breeds: the Corriedale (merino-Leicester-Lincoln); Columbia (Rambouillet-Lincoln); Targhee (Rambouillet-Lincoln); Romeldale (Romney Marsh-Rambouillet-Corriedale); and Polworth (merino-Lincoln).

SHORT-TAILED SHEEP

The northern short-tailed sheep, similar to the medium-wool type, are found primarily in Scandinavia. They account for about 3 percent of the world sheep population and are characterized by high reproduction rates. At present they are used experimentally to a limited extent in the United States for crossbreeding, primarily to increase the number of lambs born to each breeding female.

FAT-TAILED SHEEP

These sheep are so named because they can store large amounts of fat in the tail and the region of the rump. They are kept primarily for their milk-producing ability, which is higher than that of other types; their wool, however, is coarse and long and is used primarily for making carpets. Fat-tailed sheep are found mainly in the extremely arid regions of Africa, the Middle East, and Asia, and they represent about 25 percent of the world sheep population. The major breeds are the Awassi, Bakhtiari, Karakul, and Karamon. The pelts of the newborn Karakul lambs are used for the manufacture of the so-called Persian fur coats.

Haired Sheep

Domesticated sheep on the whole are much more woolly than their wild allies, but some breeds lack wool and are covered with hair. They are found primarily in tropical regions and are used primarily for meat. Among the breeds of this type are the Black-bellied Barbados, Blackhead Persian, and the Peliquey.

Production

Various systems have evolved in sheep production. In the western United States, large bands are grazed on the open range under the control of shepherds. Usually fine-wool or crossbred ewes, they are bred to rams of the Down breeds for the production of lambs for meat.

The bands may be moved from place to place to take advantage of cheap feeds and natural forage. Similar operations are used on lands with developed pastures. If abundant forage is available, the lambs may be marketed directly after weaning. If the lambs have not reached marketable condition, they may be moved to feedlots and given additional food before they are sold. Some breeders specialize in the production of purebred animals for use in the large commercial flocks.

These purebred flocks are considerably smaller in size, and genetic selection for superior productive traits is emphasized. The income from such flocks is derived primarily from the sales of the superior animals, especially the rams, to commercial breeders. In the midwestern United States, sheep enterprises may consist of a farm flock containing small numbers of animals. The flocks are maintained primarily as an auxiliary unit to utilize the by-products of other producing units, particularly grain crops.

TYPES OF WOOL

Wool is classed as follows: fine, usually short-staple wool of Merino fineness and including Delaine Merino, combable fibres 2 in. (5.1 cm) or more in length; medium, or mutton, 2 1/2 to 6 in. (6.4–15.2 cm) long, e.g., Cheviot and Southdown; long-staple, 10 to 15 in. (25.4–38.1 cm) long, loosely crimped, e.g., the Lincoln and the Cotswold; and carpet, 1 to 15 in. (2.5–38.1 cm) long, strong, coarse, and usually blended for uniformity.

For industrial purposes the fibre of the camel, Angora goat, Kashmir goat, llama, alpaca, and vicuña is classed as wool.

Sheep are sheared with mechanical clippers. The fleece thus recovered is classed as lamb's wool, or first clip; hog wool, clipped from sheep 12 to 14 months old; wether wool, from older animals; taglocks, the ragged, discoloured portion; and pulled wool, usually weakened when recovered by sweating or chemical processes from sheep slaughtered for mutton.

Alpaca

Alpaca, South American mammal closely related to the llama and vicuña. Partially domesticated, the alpaca possibly was derived from the wild guanaco. The natural habitat of the alpaca is the Andes in South America. It is most commonly kept in flocks by the Native Americans in the highlands of Peru and

Chile. The alpaca, a surefooted animal, is smaller than the llama, has longer, softer wool, and ordinarily is not used as a beast of burden. Like the llama and camel, alpacas sometimes spit. After 11 months' gestation the female gives birth to one offspring. The young have short wool until they mature.

Flocks of alpacas graze almost wild on high plateaus, 4267 to 4877 m (14,000 to 16,000 ft) above sea level. At shearing time the Native Americans drive them to the villages and clip off about 20 cm (about 8 in) of the wool. The alpaca provides white, gray, or yellow wool, although the black and dark brown fibres are especially valued. The fibre is elastic and strong and is straighter and silkier than sheep's wool. Although the flesh of the alpaca is palatable, the animal is kept primarily for its wool, from which a fine cloth is made.

Llama

Llama, common name for a long-eared South American ruminant that is domesticated from the guanaco. The llama stands 0.9 to 1.3 m (3 to 4.3 ft) high at the shoulder and is usually white, blotched with black and brown; sometimes it is pure white or pure black. The female llama gives birth to one or rarely two offspring once every two years.

Male llamas have been used as beasts of burden in the Peruvian and Bolivian Andes Mountains for more than 4000 years. The surefooted animals can carry as much as 91 kg (200 lb) for 12 hours a day, but they are not ridden. When weary or overloaded, llamas lie down and refuse to move, often spitting at their driver. The females are raised for their flesh, which tastes somewhat like mutton, and for their milk, which is used extensively in western South America. The meat of the males is tough and rarely eaten. The long, coarse wool of both sexes is used in the weaving of textiles, and the skins are tanned for leather. Llama tallow is used for making candles. The long hair is braided and used for rope, and the dried excrement is used as fuel. For related animals in the same genus.

Chordata

Chordata phylum of animals having a notochord, or dorsal stiffening rod, as the chief internal skeletal support at some stage of their development. Most chordates are vertebrates (animals with backbones), but the phylum also includes some small marine invertebrate animals.

The three features unique to chordates and found in all of them at least during early development are: the notochord, composed of gelatinous tissue and bound by a tough membrane; a tubular nerve cord (or spinal cord), located above the notochord; and gill slits leading into the pharynx, or anterior part of the digestive tract (the throat, in higher vertebrates). In addition, all have blood contained in vessels, and the tunicates and vertebrates have a ventrally located heart.

All have a postanal tail, that is, an extension beyond the anus of the notochord or backbone and of the body-wall musculature, containing no internal organs. In vertebrates—animals of the subphylum Vertebrata—a backbone of bone or cartilage segments called vertebrae develops around the notochord; its upward projections partially surround the nerve cord. In many fishes and in early fossil amphibians and reptiles the notochord persists in the adult and is enclosed by the vertebrae; in higher vertebrates, however, it disappears during embryonic development. There are two invertebrate subphyla: the Urochordata, or tunicates, and the Cephalochordata, or lancelets. A third invertebrate group, comprising the acorn worms and their relatives, shows affinities with chordates and has sometimes been considered a chordate subphylum, but is now often classified in a phylum of its own, the Hemichordata.

Tunicate

Tunicate marine animal of the phylum Chordata, which also includes the vertebrates. The adult form of most tunicates (also called urochordates) shows no resemblance to vertebrate animals, but such a resemblance is evident in the larva. The most familiar tunicates are the sea squirts, or ascidians (class Ascidiacea). Adult sea squirts are sedentary, filter-feeding, cylindrical or globular animals, usually found attached to rocks, shells, pilings, or boat bottoms.

The soft body is surrounded by a thick test, or tunic, often transparent or translucent and varying in consistency from gelatinous to leathery. The tunic (for which the tunicates are named) is secreted by the body wall of the adult animal. It is composed of cellulose, an almost unique occurrence of that material in the animal kingdom. Two siphons project from the animal's body; water enters the incurrent siphon at the top of the body and leaves the excurrent siphon at the side. Food particles are filtered from the water by the pharynx, which occupies most of the body, and are then passed into the digestive system. Some species reproduce by budding, resulting in the formation of colonies of sea squirts, joined at their bases by slender stalks or embedded in a slab of common tunic material.

In addition, nearly all species reproduce sexually and are hermaphroditic. The free-swimming larva, called a tadpole, has a muscular tail and is similar in appearance to a frog tadpole. The larva has the characteristic chordate features also found in the embryos of vertebrates: a dorsal, hollow nerve cord; a stiffening rod, or notochord; and gill slits leading into the pharynx. The tadpole eventually settles and undergoes a drastic metamorphosis into the adult form. A common solitary sea squirt of both coasts of North America is the slender, yellow, transparent *Ciona intestinalis,* about 2 in. (5 cm) tall.

The sea peach, *Tethyum pyriforme,* is a round, peach-coloured sea squirt found from Maine north. Sea grapes are clusters of the greenish colonial squirt, *Molgula manhattensis,* common from Massachusetts south. Golden stars are colonies of various *Botryllus* species; the bright yellow individual animals are grouped in starlike clusters in a flat, encrusting, greenish tunic.

Amaroucium species form colonies of minute animals embedded in a grayish, gristly tunic; chunks of such colonies, often washed ashore, are known as sea pork. There are two other groups of tunicates, both found in the plankton of open oceans. The salps (class Thaliacea) are barrel-shaped tunicates, open at both ends; they swim by muscular contractions that force water through the body. The larvaceans (class Larvacea) retain the larval form, with a tail and a notochord, as adults. A commonly held theory maintains that vertebrates evolved from animals like the larvaceans. Larvaceans have no tunics, but secrete gelatinous containers, called houses; these are used to filter food from the water and are continuously discarded and replaced. Tunicates are classified in the phylum Chordata, subphylum Urochordata.

Hemichordata

Hemichordata small phylum of marine invertebrates closely related to both the echinoderms (phylum Echinodermata) and chordates (phylum Chordata). Acorn worms, class Enteropneusta, are the most common hemichordates. The body is composed of an anterior, conical proboscis, a short collar, and a long, wormlike trunk. Gill clefts in the pharynx and a hollow nerve cord in the collar suggest relationship to chordates, while the ciliated larvae are similar to those of echinoderms. Hemichordates live in burrows or under objects in shallow water and feed on detritus. *Balanoglossus gigas* may reach a length of nearly 5 ft (1.5 m), but most species are considerably smaller. Also in the phylum is the class Pterobranchia, whose members are tiny deep-sea creatures that form colonies by secreting a collagenous exoskeleton.

They are considered more primitive than the enteropneusts. Pterobranchs have been identified with graptolites, hemichordates of the class Graptolithina (formerly believed to be extinct) since the discovery in 1992 of a living pterobranch identical to fossil graptolites.

COMPOSITION AND CHARACTERISTICS

Wool consists of the cortex, overlapping scales (sharper and more protruding than those of hair) that may expand at their free edges causing fibres to intermesh; elasticum, the inner layer; and a core. When soaked, the elasticum and core contract, shrinking the fibre. Elasticity resulting from the molecular structure of wool and resiliency from its crimp make wool fabrics crease resistant. Fine wool will stretch one third its length. Wool is warm because its fibres are nonconductors of heat and its crimp permits it to enmesh still air.

It is highly absorbent and releases moisture slowly. Its tensile strength is one fourth greater than that of cotton. A protein compound of complex chemical composition, it is soluble in hot caustic soda.

Wool Production

The wool is sorted as to fineness, crimp, length of fibre, and felting qualities. Dirt, suint (dried perspiration), and lanolin are removed by a soap-alkali scouring; by the expensive naphtha solvent method, which retains the full strength and softness of the fibre; or by freezing and shaking. Wool may be carbonized to remove vegetable matter. It is bleached and dyed as raw stock, yarn, or in the piece; it is oiled to withstand processing and is often blended.

Woolen goods are woven from carded short-staple fibres into soft yarns adapted to fulling and napping. Worsted fabrics such as whipcord, gabardine, and serge have a hard, smooth texture. Originally made only from long-staple fibres, worsted yarn is now spun also from medium or short fibres.

The fibres are carded, the resulting sliver gilled to straighten the fibres and double them for uniformity; subjected to successive combings to remove nails (short ends) and lay the fibres parallel; then drawn into roving and spun, usually by the rapid, continuous ring method, and twisted. Although the twill weave is usual for worsteds, the same weaves may be used as for woolens without the pattern being obscured by the napping, fulling, and shearing processes commonly employed in finishing woolens.

ORIGIN OF WOOL PRODUCTION

No known wild sheep are wool bearing. The supposed ancestors of the domestic sheep had long hair and a soft, downy undercoat, which under domestication gradually became wool, while the long hair disappeared. In this development, breeding, feed, climate, and protection were influential, as shown by an atavistic return of neglected sheep to long hair and rudimentary wool.

In the tombs and ruins of Egypt, Nineveh, and Babylon, in the barrows of early Britons, and among the relics of the Peruvians, fragments of woolen fabrics are found.

The Romans as early as 200 B.C. began to improve their flocks, which became the progenitors of the famed Spanish Merino sheep. The Britons kept sheep and wove wool long before the Roman invasion, but the establishment by the Romans of a factory at Winchester probably improved their methods. William the Conqueror brought into England skilled Flemish weavers.

Henry II encouraged wool industries by laws, cloth fairs, and guilds of weavers. Edward III brought weavers, dyers, and fullers from Flanders. England became the great wool-producing country of Europe, and wool was the staple of its industry until cotton began to overshadow it in the 18th cent.

In the American colonies, sheep raising started in Jamestown. Stringent English laws against exporting wool passed in an attempt to force the use of English cloth on the colonies, early drove the settlers to the raising of sheep. George Washington imported sheep and brought spinners and weavers from England. Early in the 19th cent., imported Merinos greatly improved the existing stock. Spinning and weaving were early established in New England, at first in homes, later in small factories. The first factory in America using water power to weave wool was established (1788) at Hartford, Conn., and was encouraged by tax exemption and a bounty on each yard woven.

KNITTING

Knitting, complex fabric-forming technique in which a continuous yarn is interlooped to form a cloth. This technique is used to produce about 40 percent of all fabrics in the United States. Needles are used to create each loop and draw it through a previously formed loop, thus developing a cloth from a series of interlinked, chainlike stitches.

The stitches in a finished fabric are arranged in a grid of interlaced stitches similar to that produced on a weaver's loom. The knit stitches that occur in vertical columns are called wales, and those in horizontal rows are called courses.

While the form that the knit stitch takes does not change, the manner in which the loop is drawn through previously formed loops varies.

This allows only part of the loop to be visible on the front side, resulting in an appearance and texture that is different when viewed from the back side. When a loop is drawn through from the back to the front, it is referred to as a knit or plain stitch; when drawn in the opposite manner, it is called a purl stitch.

There are two types of knitting: In weft knitting, crosswise stitches produce a fabric or garment that is looser and more elastic than that produced by warp knitting, which uses lengthwise stitches. Hand knitting—in which a person works with two long needles and a ball of yarn—is weft knitting.

Machines also do this type of knitting, often at extremely high speeds, producing sweaters, hosiery, underwear, and similar apparel. Warp knitting, however, is done only by machine and is used to produce fabric for lingerie, dresses, draperies, upholstery, and other goods. By selecting different colours and textures of yarn and by using varied stitches and knitting techniques, a multitude of patterns can be produced.

In machine knitting every stitch is made by an individual or separate needle. Thus, there must be one needle in the machine for every wale, or rib, in the fabric. Needles are held in place by a slotted piece of metal called a needle bed and are raised and lowered in the bed to perform the knitting process.

Needles of the spring-beard or latch type are designed to hold the yarn while the loop is formed, to cast it off, and then to repeat the sequence. Compared to woven fabrics of similar fibre content, knit fabrics are easier to care for, offer greater flexibility in styling, and are lower in cost, but they are more likely to lose their shape and to develop runs and holes that are difficult to repair.

Loom

Loom, machine used in weaving fabrics from yarn or other fibres. A woven fabric produced on a loom is made by interlacing two sets of threads at right angles. The longitudinal threads are called the *warp,* and the transverse threads are called *filling* in the United States and *weft* in the United Kingdom and British Commonwealth countries.

Weaving Process

Weaving on a loom is different from other methods of fabric manufacture, such as knitting, which consists of interlooping one or more threads of yarn; braiding, in which several threads of yarn are interlaced in a manner not parallel to the fabric axis; lacemaking, in which one set of threads is passed between other groups of threads; and bonding, in which a web of fibres is held together by a cementing medium.

The basic process of weaving consists of passing the filling threads alternately over and under the warp threads. The raw materials from which weaving threads are produced vary from animal fibres such as wool, mohair, camel hair, rabbit hair, alpaca hair, vicuña hair, and silk, to vegetable fibres such as cotton, and the wood and leaf fibres such as flax, hemp, jute, and ramie. In the large and continually growing area of synthetic fibres, use of fabrics such as rayon and acetate made from a natural product, primarily cellulose, is widespread.

Truly synthetic fibres, however, are more widely used than cellulose-based synthetic fibres. This group is primarily represented by, polyester, polypropylene, and acrylic fibres. Another synthetic group, spun glass, or fibreglass, is used successfully in such woven products as draperies. All the natural fibres with the exception of silk occur in finite lengths, and therefore have to be spun into thread before weaving. Silk and most of the synthetic fibres come in continuous-filament form, although cut lengths for spinning are also made in all the synthetic fibres.

Weaving fabric with a hand or power loom involves several steps. To prepare the loom for operation, the warp threads are installed in the loom and held under tension; they form a surface of closely spaced, parallel threads. To begin the cyclic process that produces the fabric, first *shedding* occurs. Shedding raises some of the warp threads so that the filling thread can be correctly placed. In a *plain weave*—a simple weave in which each filling thread is alternately threaded above and below the warp threads—every other warp thread is raised.

The space between the raised and unraised warp threads is called the *shed.* Next, in the *picking* process, a device called a *shuttle* pulls the filling thread through the *shed.* Then, in *beating up,* a device called a reed forces the filling thread against the previously placed filling threads to form a compact fabric. Finally, the raised warp threads are lowered, and a new set is raised in preparation for a new cycle; this locks the filling thread into place, above certain warp threads and below others.

Origin of Loom

The use of the hand loom dates from ancient times and was practiced by Chinese and Middle Eastern civilizations before it began in Europe. The use of the hand loom is widespread in many developing countries, and is still used in industrialized countries to make handcrafted materials such as decorator fabrics.

The hand loom is mounted on a frame that gives the necessary support to the moving parts. The warp threads are parallel to the ground. At the back of the frame is a cylinder around which the warp threads are wrapped that keeps the warp threads under tension.

Periodically, the cylinder is rotated to feed more warp thread as finished fabric is created. In the region where the filling thread is installed, each warp thread passes through a small eye set in the middle of a vertical wire called a *heddle.*

Groups of heddles are connected to a wood or metal frame called a *harness* so that the group of heddles, together with the warp threads passing through the heddle eyes, can be raised or lowered in a single operation.

In a hand loom the picking is performed by hand. The weaver passes the shuttle containing a spool of filling thread through the shed. After each passage through the shed, the filling thread, individually called a *pick,* is beaten up against the previous one by moving forward a comb of closely spaced wire teeth. After this beating up, the weaver drops the heddles that were raised and lifts the ones that were lowered, thus changing the shed.

The first move toward mechanization of the loom was the *flying shuttle,* patented by the British inventor John Kay in 1733. This consisted of a lever mechanism that drove the shuttle across the loom along a track. The flying shuttle greatly increased the speed of weaving and permitted picking to be performed by one person.

Power Looms

The first successful mechanized loom was perfected by another British inventor, Edmund Cartwright, who patented the first power loom in 1786. In subsequent years he and other engineers made certain improvements on the loom, and by the early 19th century the power loom had come into general use. The power loom essentially resembles the hand loom, but has several accessories, such as mechanisms to stop the loom if the warp or filling breaks or if the shuttle does not travel the entire distance across the loom, and a mechanism for changing shuttles without stopping the loom. At one end of these looms, a magazine containing several full spools of thread is placed; the loom has a device that rejects an empty spool and substitutes a full spool.

For fabrics more complicated than plain weaves, more sophisticated looms are required. For weaving a *dobby loom* is generally used. The difference lies in the operation of the harnesses. A two-harness loom—that is, a loom with two sets of heddles, can only produce a plain weave. Twill fabrics require looms with three or more harnesses, and other types of fabrics may need a minimum of five harnesses. In dobby looms the harnesses are operated by a *dobby head* that determines which harness is raised or lowered for each passage of the shuttle.

A further refinement is the *Jacquard loom,* perfected by the French inventor Joseph Marie Jacquard in the early 1800s. In this loom, the heddle harnesses are absent; the individual warp threads are controlled by a series of upright wires that are attached to the *Jacquard head* high above the loom. The weaving process is controlled by a series of cards with holes punched in them that correspond to the weave pattern. A space can be punched or not punched for each warp thread position.

The cards are fed through the Jacquard head at the rate of one card for each pick of filling. The presence or absence of holes in each card determines whether each warp yarn is raised or lowered. Extremely elaborate weave patterns can be produced by this system.

The reproduction of fabric designs on Jacquard cards is a highly skilled operation. For weaving fabrics with dyed threads, where shuttles of different-coloured filling threads are used, the changing of shuttles in accordance with the colour scheme of the fabric to be woven is accomplished by the use of multiple shuttle boxes, each holding a shuttle with a different-coloured thread.

MODERN LOOMS

Today, the conventional power loom is considered archaic, noisy, and generally inefficient. Various alternative looms have been developed, most of which attempt to do away with the shuttle; hence, they are known as *shuttleless looms.* Among the first was a Swiss invention that replaced the shuttle by a *dart.* Instead of carrying its own supply of thread, the dart grasps the thread from a large package and carries it through the shed. Another important type of shuttleless loom is the *fluid-jet loom,* which uses a high-pressured burst of air or water to propel the filling thread through the shed; no mechanical device is needed to direct the filling thread. Many modern fabric-producing factories use shuttleless looms because they are usually quieter and faster than conventional looms.

Hydroponics

Hydroponics, term applied to cultivation of plants in nutrient solutions without use of soil. Soilless growing of cultivated plants began in the 1930s as an outgrowth of the culture techniques used by plant physiologists in plant nutrition experiments. More recent successful methods of soilless growth differ in particulars but have two common features: (1) nutrients are supplied in liquid solutions; and (2) plants are supported by porous material, such as peat, sand, gravel, or glass wool, that acts as a "wick" in relaying the nutrient solution from its source to the roots.

Nutrients

Through photosynthesis, green plants manufacture their own organic food, using carbon dioxide and oxygen as raw materials. The nutrients usually supplied to plants by soil are almost entirely mineral salts. Plant physiologists have discovered that plants require carbon, hydrogen, oxygen, nitrogen, phosphorus, potassium, magnesium, sulfur, calcium, iron, manganese, boron, zinc, copper, and probably molybdenum.

Carbon, hydrogen, and oxygen are obtained in large quantities from water and air, but the remaining elements are ordinarily supplied as salts by the soil. The relative amount of each of these elements required for normal growth is different in each plant, but all plants require relatively large proportions of nitrogen, phosphorus, potassium, magnesium, sulfur, and calcium. Iron, manganese, boron, zinc, copper, and molybdenum are supplied in minute quantities, and are called micronutrients or trace elements.

The specific salts used to supply these elements may be varied at the discretion of the grower; a typical solution of primary minerals is composed of distilled water containing potassium nitrate, KNO_3, calcium nitrate, $Ca(NO_3)_2$, potassium acid phosphate, KH_2PO_4, and magnesium sulfate, $MgSO_4$. In solution, the salts dissociate into ions; potassium nitrate is available to plants as the ions K^+ and NO_3^-. A solution of micronutrient salts is added to the solution of primary elements to complete the nutrient solution. A small amount of fungicide is usually added to prevent the growth of mold.

Hydroponic Culture Methods

Several culture techniques are used. The most practical commercial method is subirrigation, in which plants are grown in trays filled with gravel, cinders, or other coarse materials, and periodically flooded with nutrient solution. The solution is allowed to drain off after each flooding and may be reused as long as sufficient minerals remain in it. The water-culture method is used widely for botanical experimentation.

A common type of water culture consists of glazed porcelain jars filled with solution; the plants are placed in beds of glass wool or similar material that are supported at the surface of the solution. Roots of the plants penetrate the beds and remain in the solution.

The least exact method, commonly called the slop method, is the easiest to operate. Coarse, clean sand is used in place of soil, and nutrient solution is poured on the sand in approximately equal amounts at regular intervals. A refinement of this practice is the drip method, in which a steady, slow feed of nutrient is maintained. Excess nutrient solution is allowed to drain off in both slop and drip methods.

Hydroponic culture methods are being used successfully to produce plants out of season in greenhouses and to produce plants in areas where either the soil or the climate is not suitable for the crop grown. During World War II, several U.S.

Army units successfully produced vegetables hydroponically at various over-seas bases. In the 1960s hydroponic farming developed on a commercial scale in the arid regions of the United States, particularly in Arizona, where research was also undertaken at state universities.

In other arid regions, such as the Persian Gulf and the Arab oil-producing states, hydroponic farming of tomatoes and cucumbers is under way; these countries are also researching an additional group of crops that may be grown by this method, as they have limited arable land.

9

Cotton Manufacturing

The manufacturing processes.-When the cotton finally arrives at the factory, it must undergo many long and interesting processes before it is finally turned into the forms suitable for consumers' use. The principal group processes were named in Chapter II: spinning, weaving or knitting, dyeing, and finishing, and finally the production of ready-towear goods. We shall now trace each of these processes in order.

Business organization o f the manufacturing processes. Generally each of these processes is conducted in a separate factory and under separate management. It is the exception to find one company owning all the factories necessary to turn out finished goods, although the number of such companies seems to be increasing. In most cases the spinning is done by one house, the weaving by another, the dyeing and finishing by a third, and the cutting up into ready-to wear goods by still another. In most cases the goods change ownership with each move from plant to plant, although weavers and knitters frequently send goods to dyers and finishing houses to be colored and finished at so much a yard or a pound. Sometimes goods are handled by speculators or middlemen, known as converters, who buy the goods as they come from the loom and hire other factories to dye and finish them. There are in this country several cottonmanufacturing concerns that do their own spinning, weaving, and finishing, but none, so far as the writer knows, has added cutting to its activities. Usually the woven and finished goods are turned over to some broker or jobber who in turn sells them to cutters-up or to the piece-goods trade; that is, the dry goods wholesalers and retailers the country over. Purchase of cotton by spinners.-It has already been suggested that the spinners must get their cotton from exchange brokers or from merchants, and also that in order to produce yarns of a certain quality they may find it necessary or economical to mix two or more varieties of cotton. All the year round new cotton is coming into the big markets of the world.

For example, American cottons begin to be available in November and December. Egyptian cottons come just a little later. India cottons are picked and shipped throughout the year, as are also those of Brazil, while Peruvian cottons arrive in February or March. Besides the new cottons, the warehouses contain cotton owned by farmers, merchants, or speculators, and held for increased prices or for filling previous contracts. From these various sources of supply of new and of old cotton, the spinner must get what he needs.

THE SPINNING PROCESS

The process of spinning is essentially nothing more than taking the loose, tangled fibers, drawing them into a smooth, uniform thread, and twisting the thread to give it strength. This process, formerly done by hand, is now accomplished by long rows of complicated and expensive machines in the spinning mill.

The cotton arrives at these mills in the bale. First the iron bands and the wrappings are removed that the bales may be broken into pieces. Then the cotton is loosened pretty thoroughly, after which any desired mixing of qualities or varieties is performed. The cotton is then passed through machines that clean it, remove all impurities, and reduce every lump into a fine downy mass. Next it is carded, and if of fine quality and intended for the finest goods, it is also combed. From the carding and combing machines the cotton passes into drawing frames, machines that begin to draw the cotton out into a thread. It is then ready for the spinning machines where the final twisting of the yarn takes place. In an up-to-date plant human hands scarcely touch the cotton from the time it enters the bale breaker until it comes out from the spinning machines a finished yarn.

The steps in the process.-The first machine, known as the bale breaker, takes the big lumps of hard-packed cotton as it is found after the wrappings have been removed from the bale, and breaks them into smaller pieces, which pass into another machine called the cotton opener, there to be torn into still finer portions. In some mills the bales are broken up by hand, instead of by the machine bale breaker. The lumps made by hand-breaking are fed directly into the cotton openers.

In the Southern States, in sections where a considerable amount of loose cotton is hauled directly from the gin to the mill without baling, there is of course no use for the bale-breaking and cotton-opening machines. These machines simply undo what the baler or compress has done for the cotton fibers.

Mixing.-After the cotton bales have been broken into small pieces, the next step is mixing. This is generally done by conveying the cotton from the bale breakers and openers to bins, where a layer of one kind of cotton is covered by another, and so on until all the cotton of a certain batch or "mixing," as it is called, has been deposited in the right proportions. At the end of the operation the different varieties or classes of cotton lie in the same bin in horizontal layers. When the mass is removed from the bin, it is taken out in vertical, or up-and-down, sections.

The picker.-From the mixing bin the cotton is conveyed by hand, or by machine in large modern plants, into a contrivance known as the picker. This pulls the cotton fibers into loose masses and delivers them in a flat sheet which looks like cotton batting, but which in cotton-mill language is called the lap.

The lap is immediately sent on to other machines known as the intermediate picker and finisher picker, each of which pulls the fibers into a still looser mass, and beats out sand, dust, and most of the other foreign matter likely to be found in a cotton bale. Each machine delivers the cotton in the form of a lap. In this form the cotton is much easier to handle and to feed into the next machine than loose bulk cotton would be. If the machines are not close together, the lap is wound into a roll like cotton batting, from which it is unwound when passed into the next machine.

The scutcher.-If there is considerable dirt in the cotton it is generally run through another series of machines known as scutclaers. In the scutchers the cotton lap gets a great deal of beating and shaking which removes the foreign matter. The cleaner the cotton is at the start, the less beating or scutching it needs. This is an important matter, for beating is sure to weaken a certain amount of the fiber, and the more beating it gets, the weaker will be the final product, the yarn.

Carding.-From the pickers or the scutchers, as the case may be, the cotton lap is transferred into the carding machine, the purpose of which is to remove all dirt, sticks, particles of leaves, and other impurities that were not removed by the pickers and scutchers. It disentangles the fibers still more, and lays them approximately parallel. Drawing out the lap into a thin filmy layer of cotton, usually about forty inches wide, it then contracts this layer into a light, round cotton rope or ribbon about an inch in diameter. This rope is called the card sliver.

The carding machine or carding engine consists essentially of two surfaces both covered with great numbers of short, sharp teeth made of wire. These surfaces, called the cards, face each other with only a narrow space clear be tween the teeth. The cards move in such a way as to brush the cotton lint as it passes between them, combing out the fibers and catching all irregularities, such as dirt, parts of seeds, short fibers, and neps (little bunches of cotton fiber that have not been loosened). The lower card is in all modern machines a large revolving cylinder, while the upper card is shaped like a cover or roof over the upper part of the cylinder. A few years ago the upper card was a stationary device, but it has now been improved so that it moves also. This has been accomplished by fastening the carding surface to an endless belt or carrier, passing it around cylinders at the two ends and at the middle, and putting in devices to keep the lower side from sagging down upon the lower card. Thus both upper and lower cards revolve, but the cylinder turns at a much greater speed than does the top card, known as the flats; the cotton is therefore brushed and combed by the flats while being carried along on the teeth of the cylinder. There are also in the carding engine other mechanical devices called respectively the lickey-in and the doffer. The first draws the cotton lap into the cards; the second removes it therefrom. The cotton is removed from the doffer by a comb. The machinery having been so adjusted that the cotton travels faster in the cards than the rate at which the lap comes in, the stream or sheet of cotton lint becomes much thinner than the lap from the scutcher that enters the carding engine.

From the doff er the thin cotton lap is carried to what is known as a trumpet mouth in which the entire lap is condensed into a round rope, three or four inches in diameter, called the card sliver. The sliver is immediately passed between a pair of calendar rollers which condense it to about the size of a broomstick. Thence it is conveyed to a little device that winds it spirally into a deep sheet-iron or tin can called the sliver can. When full, this can is carried to the drawing machinery. In case yarn of a very high count, or of a very fine size, is to be made, the sliver is taken to the combing machines to be combed before it is sent to the drawing frame. Practically all sea-island cotton, all yarn of sizes from 60's up, and certain fine yarns of lower counts are thus not only carded but combed also. In some cases, especially in producing yarn for medium-priced cotton underwear and warp for velvets and fancy woven fabrics, the cotton lint is run through the carding machine a second time instead of through a combing machine. The latter process has been found to be cheaper than combing.

Combing.-The combing machine simply does more perfectly what the carding machine starts to do in the way of cleaning the fibers, removing the neps and short fibers, and getting all the remaining fibers to lie smoothly and in parallel order.

The usual combing method consists in taking several cans full of sliver to the machine and starting eight or ten, or even more, into the combing machine at once. These slivers are first converted into a smooth lap about nine inches wide and then are passed as a single stream or lap into the combing device. Here by means of rollers, nippers, and rows of metal teeth the fibers are thoroughly combed, and all short fibers, any dirt remaining, and other foreign matter are pushed aside as waste. There has been great improvement in cotton combing during recent years. The combing machines now utilize cotton fiber that is not more than seven-eighths of an inch long, and yet produce a sliver that is almost as fine as that from the best cotton a dozen years ago.

Drawing.-Next, the slivers from the carding engine or from the combing machine are taken to the drawing frame. About six slivers are introduced together. The drawing frame takes these six and so draws them out that the resulting sliver is no larger than one of the card slivers. Drawing simply mixes the cotton, causes the fibers to arrange themselves in the best manner possible for the spinning that is to follow, and draws the combined six slivers out into the thickness of one. The drawing frame accomplishes this by means of several sets of rollers through which the slivers pass, each following pair of rollers having a higher rate of speed than the pair preceding. Requirements differ, but most cotton is run through drawing frames three times.

Next the cans full of sliver from the drawing frames are transferred to the fly frames. Here the cotton is drawn out into still smaller diameters and given the small amount of twist sufficient to allow the sliver to be drawn out further without breaking. For very coarse yarns the cotton passes through but two fly frames, the ones known as the slubbing frame and the roving frame, the product from the slubbing frame being called slubbing or slub, and the product from the roving frame roving. For medium yarns the cotton passes through three fly frames, slubbing, intermediate and roving, while for very fine yarns it is passed on to a jack frame to be drawn out still more. The cotton from the fly frames, now called roving, is wound on spools or bobbins and taken to the spinning machines.

Comparison of the mule and ring frames.-Spinning completes the drawing out of the cotton roving to the required size, and gives it the proper amount of twist. There are two kinds of spinning machines in use, one known as the mule frame and the other as the ring frame. The mule frame is the older machine invented in England, while the ring frame is an American invention dating back to 1828. Each has certain advantages. For example, mule spinning produces a soft, oozy, elastic yarn that is very satisfactory for hosiery, underwear, and especially for woolen goods. The mule frame is more common in the woolen-spinning industry than in the cotton. Although it is much more complicated than the ring frame, takes up more space, requires more skillful operators, and does not run so fast as the ring frame, yet it is not so hard on the yarn, and it produces yarn of superior quality. The ring frame is not usually used for yarns running above 60's. The mule frame has three distinct and separate motions: first, the roving is drawn out; second, it is given the necessary twist; and third, the twisted yarn

is wound on the spindle. The ring frame draws, twists, and winds, all at the same time. In both cases the yarn is wound on bobbins, cops, or tubes fastened on the spindle. This is the form in which it usually leaves the spinning mill. Sometimes, however, it is rewound on wooden spools by another machine called the spooler. Cotton yarn is quoted on the market as so much a pound for a certain size on cops, cones, or tubes, or in skeins. Often the yarns are combined, twisted double or triple-or two-ply or three-ply, as it is called. Special machines perform this operation.

The term *"spinning"* is sometimes used to denote this final process in the production of the yarn. This involves attenuating (stretching) the yarn to the required tex. Giving the thread strength by adding twist. And winding it on to a bobbin. There are two main methods:

- Mule Spinning
- Ring Spinning

The MULE was originally developed by Samuel Crompton from the "jenny". He never patented his invention and this must have helped its wide introduction. The mule operated in two stages. In one stage the whole 'front' of the machine (perhaps 100 feet long) moved away from the back part stretching and twisting the thread as it did so. It would move several feet (say 5 feet). In stage two the front carriage moved back and at the same time wound the stretched yarn on to a bobbin (or *cop*). With the early mules the carriage was moved forward by the operators turning a wheel but the invention of the self-acting mule meant that the carriage moved forward itself. Mules would be placed in lines so that the front of one faced the front of the next. As the carriages moved forward, towards each other, only a narrow gap would be left between them for the spinner to walk between. The mules were tended by spinners, piecers, doffers. Piecers would mend broken threads and doffers would remove the full cops. Often they would be men. Mules could at one time produce much finer yarns than ring frames but as the latter have become more capable the mule has become less used.

Fig. Self Acting Mule

RING-SPINNING was a development from Richard Arkwright's "water frame". Many Lancashire manufacturer's found that Arkwright's patents were too restrictive so the early water frames lost out to the mule. Ring spinning was not used much in Lancashire until the later years of the 19th century (though invented in USA in 1829) and is carried out on a machine called a ring frame.

The process is continuous rather than intermittent, and higher speeds can be achieved. This became the dominant method by mid-1900s. Ring frame tenters were often women and again the full bobbins are removed by doffers. THROSTLE spinning was also developed from the water frame.

Fig. Ring Spinning Frame

This is as good a point as any to say something about this occupation which can have a number of meanings. The most usual meaning in Lancashire census records is someone who looks after something, cares for something, tends it. So an "engine tenter" is someone who looks after an engine. Another usage which goes back to the early days refers to the days when bleaching was carried out by putting the cloth in the sun. In small quantities the cloth could simply be laid on the grass but later long lines were set up in bleaching crofts and the cloth was hung from these lines on "tenterhooks". The workers who did this could be "crofters" or "tenters" or "tenterers".

So prehaps when modern chemistry enabled bleaching to be done inside the factory the word "croft" stuck. Perhaps the term "crofters" was also applied to these inside workers. The yarn which emerges from the spinning process cannot usually be woven directly and needs some preparation. Winding is the process of transferring the yarn to larger bobbins or cones. The idea is to get a long continuous length. Weft-winding inolves winding on to smaller bobbins that will go into a shuttle. "Winder" is a common occupation in the census records. The beam is a long cylinder with flanges and perhaps 600 threads are wound on to it side-by-side. The machine is watched over by a "beamer". The full beam is very heavy. In early days beaming was often done in the weaving mill but then tended to be transferred to the spinning mill which would send the full beams to the weavers.

Note that this is more specifically called a "warper's beam" The yarn is a little fragile for the rough treatment imposed by the weaving process and a "size" is applied to make it more robust. A number of warper's beams (as above) are placed at the back of the sizing machine and the yarn is drawn through and wound on to a "weaver's beam". If the machine is fed by 8 warper's beams of 500 threads each then the weaver's beam will have 4000 parallel threads. Generally the set of warper's beams will produce up to 20 weaver's beams each of 1000 yards or more. The operative is called a "tape sizer" or a "taper". This was a skilled job to get the right degree of dryness.

YARN TECHNOLOGY

Warp and weft yarns.-For purposes of making cloth two classes of yarn are spun, namely, a loose, slightly twisted yarn for the filling, not requiring much strength, and a hard twisted, strong yarn for the warp. Each class is sold as such by the spinners to weavers, and the weavers insist on getting the proper qualities in each to suit their purposes.

Long fiber for warp, short fiber for weft.-In cotton spinning it is profitable to use the longer cotton fibers for the warp and the shorter fibers for the weft or filling. Some of the shortest varieties of cotton, such as Mobile, are used only for filling. The reason for this is that the warp must be spun harder and made stronger than the filling. The warp must undergo much handling and strain. It must be threaded through the heddles of the loom and stretched across the loom. Every move of the heddles up and down subjects it to new strains. Weft, on the other hand, is simply wound onto the bobbins that go into the weaving shuttles, and unwound into place between the warp threads. At no time need it be much stronger than to support its own weight; hence cotton can be spun into a much finer yarn or thread if it is to be used as weft or filling than if it is to be used as warp. Whenever warp and weft yarns of the same size are used, the warp is usually made from a longer fibered higher grade cotton than the weft. The shorter the fiber, the coarser the yarn must be made to have requisite strength, but weft, or filling, need not be so strong, and therefore not so coarse as warp yarn.

Cotton yarn sizes.-Cotton yarns used in the textile industry go by numbers, the number depending upon how many yards are required to make one pound. The unit is 84o yards; that is, yarn of size I, which is taken as the unit or standard, runs 840 yards to the pound. Yarn size 10 would contain 8,400 yards to the pound, and yarn size 100 would contain 84,000 yards to the pound. The spun cotton is reeled into skeins, and these are tied together into hanks. A regular spinner's skein contains 120 yards, and since there are seven skeins in a hank, the hank contains 840 yards. So when a hank weighs one pound it is called size z, or simply Vs. If the hank weighs half a pound, the size is 2's If the hank weighs only an ounce, 1/16 of a pound, the size of the yarn is Ws, and so on, the numbers running up as high as Zoo's, which means 168,000 yards to the pound.

What a pound of cotton contains.-It is surprising to most students to learn for the first time that one pound of cotton fibers can be turned into so much yarn. When yarns of the proper sizes are used in making fabrics, one pound of cotton produces one and one-half yards of denim. It produces four yards of sheeting, four yards of bleached muslin, six yards of gingham, or seven yards of calico. It is even sufficient for ten yards of lawn, twenty-five hand kerchiefs, or fifty-six spools of No. 40 sewing cotton. But one ceases to wonder at this when he learns that there are 140,000,000 fibers in a pound of upland cotton, and that if these were placed end to end, the line would extend a distance of over 2,200 miles.

Use of yarn numbers in handling cotton.-In speaking of the quality of a given kind of raw cotton, it is customary among manufacturers to state what size of yarn it will spin. For example, in speaking of India cotton, it might be stated that it will spin 12's This means that its fiber is so short that it will not make a yarn finer than number 12, which runs 10,080 yards to the pound-a very coarse cotton yarn, such as is used in heavy sheeting, drills, and denims. Sea-island cotton will make a yarn as fine as 200's and it has even been claimed that it has been drawn out as fine as 400's It can well be imagined that cloth made of such fine yarn, even though closely woven, would be like gossamer, as light as air. Between those two extremes occur all the sizes of yarns used in ordinary fabrics.

The numbers spun from different varieties.-The following are some of the leading varieties of cotton with the average sizes of yarn made from them and some examples of uses made of the yarns. There is considerable variation in each kind of cotton and the table aims to present simply a general idea of the more frequent cotton qualities. Remember that the filling or weft yarns can be made from a fourth to a half finer than those for the warp.

Mixing cottons.-When manufacturers desire to produce a fabric that requires a certain size of yarn, they do not always buy merely the one kind of cotton, which, according to the table, will make that size. It is not always possible to get a full supply of any one of the kinds named; furthermore the list contains only a few of the many varieties that come to market, each with its own special qualities. There may be a crop failure in some one variety of cotton, or perhaps there is strong competition for this variety, making the price too high for ordinary use. In order to produce yarn of a certain size a manufacturer usually buys part of his supply of cotton of better grade than necessary, and part of poorer. These he mixes in such proportions as will give just the grade desired.

Sometimes as many as six or eight kinds of cotton are mixed in order to get the right quality in a yarn. Consequently, mixing cottons is a science in itself, and requisite not only in order to get a yarn of a certain size, but also to get the right smoothness, fineness, strength, and glossiness, or the right feeling to the sense of touch. Those who manufacture certain well-known brands of goods desire to maintain exactly the same qualities in those goods year after year. But all of the cottons naturally vary from year to year according to the season. The same variety will not have exactly the same qualities every year. To insure fixed qualities in his standard goods, the manufacturer employs blends or mixtures that will give the same result every time. The proportions of the various kinds of cottons are varied as may be necessary to preserve the same qualities in the mixture.

THE WEAVING PROCESS

The steps in the process.-Preliminary to weaving is, of course, the determination of the kind of cloth to be made and the amount of yarn going into it. By arithmetical computations, it is found how many warp threads will be required to make the cloth of the required width. This is spoken of as finding the required number of "ends." The filling or weft threads are called "picks."

Winding the warp.-When the number of ends needed has been computed, arrangements are made to have these wound on the loom warp beams in the necessary manner. The threads are distributed along the beam at equal distances and wound on very evenly and carefully, each at the same tension, for this even tension in all of the warp threads on the loom is highly desirable during the weaving.

Sizing.-After the warp is strong and smooth, the warp beams with the warp wound thereon are placed in the looms, and weaving is begun. But most warp needs further treatment before it is thus ready to be used in weaving, especially if it is a single-twist yarn and not very strong. It must first be sized, that is, given a coating of some stiffening substance that will give it strength, stiffness, and smoothness. Without this sizing it would be liable to break when subjected to the strain and friction of weaving. Sizing is composed of many different substances-in most cases, of some kind of starchy substance, such as wheat flour, cornstarch, potato starch, or sago flour. This substance is usually boiled in water, and a little tallow, paraffin, or other oily substance added to soften it. 1'o this some preservative is added, such as chloride of zinc, cresol, or salicylic acid, to prevent souring, mildew, or other bacterial action or vegetable growth. The sizing applied to the warp has a purpose distinct from that used in finishing cloths. The latter will be described later.

The method of applying the sizing to the warp yarn is, somewhat as follows: The warp beams holding the warp yarn are taken to the tank which contains the sizing. The tank has a roller in it with the lower side immersed. The ends of the warp from the warp beam are passed under this of cotton fibers. This is especially true around the principal machines. This waste, known as "soft waste" until the fiber has left the spinning machines, and thereafter as "hard waste" (the particles of yarn or of cloth in the later processes), is carefully collected and used again in making cheaper yarns and goods. A good deal of the waste from the American cotton mills has in the past been exported, especially to Germany, where machines have been invented to take care of such matter and use it to better advantage than can our machines. It is used, for instance, in making cheap cotton goods for home trade among the German poor, and for export trade with China and other parts of the world where flimsy cottons are in demand. It has found a place mainly, however, as filling for cotton blankets, catton flannels, cheap trouserings, towels, carpets, mats, sacks, lamp wicks, wadding, twine, etc. Lately American manufacturers have begun to install machines in the mills in this country to use the waste and to work it up at home. The utilization of cotton soft wastes will no doubt in time constitute an important part of our cotton industry.

Cloth inspection and repairing.-After the cloth has finally come off the looms, it must be inspected for any weak places, tears, or other imperfections. These are repaired as well as possible and the loose ends of warp and weft trimmed off, a process known as bztirling. Next the cloth is suitably marked if it is to be sent to bleacheries or dyeworks, and finally it is measured, rolled into bolts, and packed in paper. The bolts are then put into wooden boxes or cases. The cloth is now ready for the finishing processes.

THE MECHANISM OF END BREAKAGE IN RING SPINNING

One basic way to increase profit and quality in the ring spinning process is to keep the end breakage rate to a minimum level. The end breakage rate is a critical spinning parameter that not only affects the maximum spindle speed but may also indicate the quality of yarn, the mechanical condition of the machine and the quality of raw materials. Therefore, it is an important parameter which determines the overall working of a spinning mill. Any attempt to explain the end breakage mechanism in terms of qualitative approach will provide an aid to understanding the problem and to tackling it to any possible degree. Substantial research works have been carried out to predict likely end breakage rates in ring spinning [1-5]. Most of these are based on large numbers of simplifying assumptions and complexity of calculations. Moreover, in many cases, they do not reflect the results which are compatible with the actual situation. This paper presents a simple model to predict the likely end breakage rates based on the statistical analysis.

Yarn breakage mechanism during tensile testing

The yarn may be assumed to be composed of successive links of fibre bundle. The length of any given link is poorly defined, because the elements of fibre bundle are not separate entities but each gradually merges into its neighbour at either end. Nevertheless, an average element length may be considered to exist. It has been reported by Lord [6] that for the carded cotton yarn the average length of the link is approximately around 3 mm. In fact, break occurs at the weakest link of the yarn, and the yarn strength is nothing but the strength of its weakest link [7]. When a yarn is subjected to a tensile test, the tension is generated in the yarn by an applied load, which is opposed by the resistance of the material. As generated tension acts on the helical structure, the yarn is compressed radially and a force normal to the fibre axis is developed. An increment in the normal force between fibre surfaces in the yarn causes an increase of resistance necessary to prevent fibre slippage. During the tensile test, the tension generated by the applied strain continues to increase.

The increase in generated tension is limited by the tensile strength of the fibres. Initially, as the extension increases, few weaker fibres with low extension may break at the weakest link of yarn. The structure is readjusted, and some of these broken fibres may continue to share the load again if the broken segments are long enough to be gripped again. The load will continue to rise depending upon the load-bearing capacity of surviving fibres. As it is further extended, more fibre rupture at the place of the weakest zone takes place. The load-bearing capacity of the surviving fibres becomes less than previously, and the recorded load starts declining. The stress on the surviving fibres concentrates, causing the rupture to propagate faster across the cross-section. The number of broken fibres becomes large enough to affect the structural integrity to such an extent that the unbroken fibres start slipping as well. The yarn rupture continues in a mixed mode of fibre breakage and slippage until it completely ruptures.

Fig. Cotton Spinning Ring Frame

End breakage mechanism in ring spinning

The mechanism of end breakage in the ring frame is significantly different from the failure mechanism of yarn in a tensile tester. In ring spinning, the end breakage occurs due to the imbalance in the tension imposed on the yarn and the yarn strength at the weakest portion. It is an observed fact that almost all end breaks in the ring frame take place just after the delivery from the front nip in the spinning zone, i.e., between the front rollers. nip and the thread guide. Therefore, an end will break when the spinning tension exceeds the strength of the weakest portion of the yarn in the spinning zone [9]. The end breakage phenomenon in ring spinning is absolutely slippage-dominated, i.e., there is no evidence of fibre breakage. The strength of yarn at the spinning zone is significantly less than the yarn strength obtained by a tensile tester. In general, the spinning tension is considerably greater than one-third of the single thread strength [1]. In fact, a very thin portion of yarn just after the delivery from the front nip causes an end breakage in ring spinning.

Assumptions

In order to derive a mathematical equation to predict the end breakage rate in ring spinning, the following assumptions have been made:

- The yarn is composed of a successive chain of links with different lengths, and the average length of links is $l\ h$. For carded cotton yarn, $l\ h = 3$ mm.
- An end break occurs when the spinning tension exceeds the strength of the yarn in the spinning zone.
- The strengths of the links of fibre bundle just issuing from the front rollers are independent variables and follow a normal distribution, i.e., a link of high strength may immediately follow one of low strength. The strength is constant within a link but varies from link to link. The strength of a link depends on the number of fibres in the link and the coefficient of variation of the strength of all the links. The coefficient of variation of the strength of the links is equal to the coefficient of variation of mass per 8 mm length of the yarn (Uster CV%). This is based on the fact that the variance-length curves of fibre assemblies run practically horizontal for cut lengths between 1 and 10 mm [10]. IV. The strength of the yarn in the spinning zone is about one-third of the mean strength of the yarn.

Materials and Methods

For the experiments, Tex carded cotton yarn was spun at a spindle speed of 16500 rpm. The spinning tension was measured using a Rothschild tensiometer-2000. This instrument works on the principle of the differential capacitor. Two fixed pins of the tension probe were removed while measuring the spinning tension to avoid twist blockage and consequently yarn breaks. The tension probe was positioned between the front rollers and the yarn thread guide. When the yarn passes over the movable pin, the pressure applied by the yarn causes a change in the voltage which is proportional to the yarn tension.

The instrument was interfaced with a computer. The output from the transducer was amplified and processed by the computer using software. A response time (the time interval between successive signals) of 1 ms was selected. Thus it was possible to observe the variation of spinning tension in one revolution of the traveller. The mean breaking force of the yarn was measured using a standard tensile tester. The CV% of the yarn mass was tested with a Uster Evenness tester II. Using the tension measuring system, the tension parametersregarding the average spinning tension and its CV% and the mean value of peak spinning tensions were obtained. The parameters for predicting end breakage rates in ring spinning.

Parameters, (symbols), units	value
Yarn delivery speed, (S), m/min	20
Average length of the links of fibre bundles ($\bar{l}_h$), mm	3
Average time taken to pass a link through the front rollers (t_1), milliseconds	9
Average time required for each rotation of traveller, (t_2), milliseconds	4
Mean breaking force of yarn, cN	375
The coefficient of variation of mass per unit length of yarn, (V), percentage	13.1
Average spinning tension, cN	32
The coefficient of variation of spinning tension, percentage	12.8
Mean value of the peak spinning tension, (T), cN	40

Each rotation of the traveller introduces five peaks to the spinning tension. But in the present study it was possible to obtain only one peak for each rotation of traveller due to the limitation of the measuring system.s response time. It is visible in Table 1 that the average time taken to pass a link through the front rollers is greater than the average time required for each rotation of the traveller. Therefore, each link while passing through the front rollers undergoes at least some peak tensions. From our assumption, the mean breaking force of the link just issuing from the front rollers, which is equal to one-third of the single yarn strength, is of the order of 125 cN.

Taking the strengths of these links to follow a normal distribution, using Equation 4, the standard deviation (ó) value would be 16.34 cN. For a break to occur, the strength of the links, which are lower than the value of the peak spinning tension, the standard normal deviation(Z) would come to 5.2. Therefore, to fulfil the condition of end breaks, the strength of the links which are lower than the value of the peak spinning tension would lie at a distance of 5.2ó from the left.

Thus using Equations 7 and 8, the probability (P) of the strength of the links lower than the mean value of the peak tension can be solved, which gives the value of $P = 1 \times 10^{"7}$. This would mean that while passing through the front rollers. nip, one link out of 10^7 links is the weakest one that causes an end to break. In other words, for every spindle, a break would repeat after a delivery of 30,000 m length of yarn, i.e., after 25 hours excluding stoppage time. Using

Equations 2 and 3, the predicted values of end breakage rates per 1000 spindle/ h would work out to 40. This predicted value of end breakage rates is approximately the standard norm of the end breaks per 1000 spindle per hour for 30 Tex carded cotton yarn under the machine.s normal running conditions.

Possibilities of Reducing end Breaks in Ring Spinning

To reduce end breaks, the following aspects should be taken into consideration: I. Since end breakage in ring spinning is related to slippage of fibres at the spinning triangle as a result of peaks occurring in the spinning tension fibre, the grip at the front drafting rollers should be increased by having a higher top roller pressure. The use of softer cots also enhances the grip at the front rollers. If the total pressure on the rollers cannot be increased, the grip at the front rollers. nip can be improved by reducing the width of the cots. II. A reduction in friction between ring and traveller could reduce the peak tension during the rotation of the traveller. III. Measures should be taken to reduce the mass irregularity of yarn straight after carding. IV. The width of the drafted ribbon at the front roller nip should be reduced. side of the mean (μ) in the normal distribution curve.

COTTON YARN PRODUCTION

The Demand for cotton yarn in Ethiopia is currently met through domestic production though the country was importing some amount in the past to fill the demand gap. The local producers are two spinning plants: Adei Ababa Yarn Factory and Edget Yarn and Sewing Thread Factory: and integrated textile mills: Akaki, Awassa, Kombolcha, Almeda and Dire Dawa Textile Factories.

Adei Ababa Yarn Factory which was established in year 1961 in Addis Ababa, has installed capacity of 4967 ton/year, while Edget Yarn and Sewing Factory has an installed capacity of 145 tonnes/year. The integrated mills produce yarn mainly for the consumption of their own wearing departments to produce fabrics, while the spinning factories produce exclusively for the hand loom sector and other consumers of marketed yarn.

In addition to the above, cotton yarn is spun at household level, and according to a textile sector study conducted by IPS, annual production of home-spun yarn is estimated at 200 ton. Traditional weavers (the handloom sector) use industrially produced yarn for warp and home – spun yarn for weft to produce traditional garments (buluko, Gabi, etc.)

PROJECTED DEMAND

Cotton yarn consumption had exhibited an average annual growth of 2 % during the years1995-to 2004. Assuming this growth trend will increase to 5% the future demand for cotton yarn is projected to range from 6,000 tonnes in year 2006 to 12,000 tonnes by the year 2020.

PLANT CAPACITY AND PRODUCTION PROGRAMME

Plant Capacity

On the basis of demand projection made for the product and in consideration of high capital costs associated with high volume of production, a plant with a capacity of 1,500 tonnes of cotton yarn per annum is envisaged in this project profile. For an average yarn count of 25, (*i.e.* for counts 10, 21 and 40 demanded in the market), the number of spindles matching this annual output is estimated at 20,000. Production capacity is based on a schedule of 300 working days per annum and 3 shifts of eight hours each per day.

Production Programme

The envisaged production programme. The schedule is worked out in consideration of the time required for gradual build-up in labour productivity and fine-tuning of machinery. Production starts at 75% of plant capacity in the first year of operation and reaches full-gear in the 3rd year of operation and thenafter.

MATERIALS AND INPUTS

The major raw material needed for the manufacture of cotton yarn is ginned cotton and dye staff. All of these raw materials could be obtained from local suppliers; additionally, there is a very high potential to grow cotton in the region. Different packaging materials are also needed in the production of cotton yarn; namely; paper cones, plastic sheets and cartonnes. Studies indicate that the combined cost of these materials is roughly about 3% of the cost of cotton used. Raw material requirement and the corresponding cost for the envisaged capacity of 1,500 tonnes of cotton yarn.

Electricity and water are the two major utilities required by the plant. Total cost of major utility items at full operation capacity of the plant is Birr 3,127,275.

TECHNOLOGY AND ENGINEERING

Production Process

Textile mills purchase cotton and receive the bales from gin yards or cotton warehouses. These mills start with raw bales of cotton and process them in stages until they produce yarn (fibers twisted into threads used in weaving or knitting) or cloth (fabric or material constructed from weaving or knitting).

The first stage in yarn making is done in the opening room. Here, bales are opened and laid in a line on the floor, side by side, near a cotton-opening machine. This machine travels along the line of opened bales, pulling fibers to be sent to a mixing machine and then on to the carding system.

Carding is the process of pulling the fibers into parallel alignment to form a thin web. High-speed electronic equipment with wire-toothed rollers performs this task. The web of fibers is eventually condensed into a continuous, untwisted, rope-like strand called a sliver. These slivers then continue to a combing machine. Here, the fibers shorter than half-inch and impurities are removed from the cotton. This process makes the sliver smoother so more uniform yarns can be produced. The drawing or pulling of this sliver is next.

The sliver is drawn out to a thinner strand and given a slight twist to improve strength, and then wound on bobbins (spools wound with the thread-like product for storage). Having completed this process, it is now called roving. The roving bobbins are now ready for the spinning process.

Spinning and dyeing or bleaching is the last process in yarn manufacturing. Modern mills draw and twist the roving into yarn and place it on bobbins. They do this quite efficiently. With the use of automatic winding, the yarn bobbins are transferred to larger bobbins called cheese cones. These cheese cones can be stored until they are needed in the weaving process.

ENGINEERING

Land, Building And Civil Works

Total land requirement of the project is estimated at 5000m^2, out of which 3000m^2 is a built-up area. Cost of building construction, at a unit cost of Birr 1,500 per m^2, is estimated at Birr 4.5 million. Total land lease cost, for a period of 80 years with cost of Birr 9.78 per m^2, is estimated at Birr 48,900. The total investment cost for land, building and civil works, assuming that the total and lease cost will be paid in advance is estimated at Birr 4,548,9000.

Proposed Location

The resource potential study of the Dire Dawa Provisional Administration indicates that cotton is one of the potential crops for the area that could be cultivated in irrigable areas of the Administration.

In view of this, the Dire Dawa is believed to be the most optimal locations for a cotton yarn factory as the above-mentioned areas have relatively better infrastructure.

MANPOWER AND TRAINING REQUIREMENTS

A spinning mill is a very labour intensive factory. The list of manpower required and the estimated annual labour cost. Total manpower requirement, including skilled and unskilled labour, is estimated at 100 persons. Correspondingly, total annual labour cost, including fringe benefits, is estimated at Birr 1,027,500.00.

The technology of cotton spinning is not new to Ethiopia as there are a number of textile factories in the country. On account of this, it is believed that some trained labour will be in the market. Hence, an on-site training programme is believed to be sufficient; this should be arranged for key production, maintenance and quality control personnel in consultation with the machinery and technology supplier. The training can be best carried out during commissioning and performance testing of the factory. Cost of an on-site training of this nature is estimated at Birr 400,000.

CARD ROOM

1. Roving Can—receptacle to hold the sliver. After it is filled it is transferred to either ribbon lap machine or drawing frame.
2. Cylinder of the card. The cotton is on this cylinder in the form of a web.

Carding Machine

When the lap of cotton leaves the picker it goes to the carding machine, where it is combed into parallel fibers by means of a revolving cylinder covered with wire teeth called card clothing. As the cotton is fed to the card in the form of a sheet or lap from the picker, it is supposed to have been freed from a considerable quantity of sand, seed, etc., but there still remain nep, fine leaf, and short fibers, which are removed during carding.

On leaving the card cylinder the lap has become a gossamer-like web thirty-nine inches broad. This web next passes through small "eyes," which condense it into a narrow band about an inch in width, known as card sliver.

When a lap is delivered from the finisher picker, it should weigh a given number of ounces per yard. The method of ascertaining the weight is to make each lap a standard number of yards in length and weigh each lap. The machine can be regulated so as to give the desired weight per yard.

Combing

When an extremely fine and strong yarn is required, in addition to carding, the fibers are also subjected to the process of "combing." This may be said to be merely a continuation of the carding process to a more perfect degree. The chief object is to extract all fibers below a certain required length, and cast them aside as "waste." This is done in order to secure the very best fibers calculated to give the strongest and best results in the spun yarn.

The process of combing follows carding. The card delivers the cotton in the form of a sliver or strand, while the combing machine requires the fibers to be delivered to it in the form of sheets, nine to twelve inches wide. This is done by taking a number of card slivers and forming a lap of them by passing the sliver through a sliver lap machine. The laps are passed through the comber. This machine consists essentially of a series of rollers, nippers, and rows of metal teeth. By the action of these, the short fibers are separated and combed out, and the long ones arranged in parallel order in the form of a thin, silky strand, in which condition it is sent to the drawing frames to be drawn out. Of course it must be understood that a combing machine is used by only a small percentage of cotton spinners. For ordinary purposes a sufficiently good quality can be made without a comber. As there is from 15 to 35 per cent waste to this operation it may be readily seen that it is costly, and limited entirely to the production of the very best and finest yarns, such as those intended for sewing or machine thread, fine hosiery, lace curtains, underwear, imitation silks, and fine grades of white goods. There are combing machines that comb short staple cotton.

Drawing

The cans containing the slivers are taken from the card or combing machine (as the case may be) to the drawing frame. The object of this machine is mainly to equalize the slivers, combining a number of them together so as to distribute the fibers uniformly.

The condition of the fibers on leaving the card or comb is such that a slight pull will lay them perfectly straight or parallel, and this pull is given by the drawing frame rollers. Of course the fibers coming from the comb are parallel, but it is necessary to alternate them by the drawing. The drawing frame is a machine consisting of a number of sets of rollers, the front roller having a greater speed than the rear ones.

COTTON COMB ROOM

The cotton in the form of a "lap" ready to pass through the comb.

The slivers, which are as nearly as possible the same weight per yard, are combined together in the drawing and emerge from the pair of front rollers as one sliver weighing the same number of grains per yard as a single sliver fed up at the back. This process is repeated two or three times, according to requirements, the material then being referred to as having passed through so many "heads" of drawing. It is not unusual to pass Indian and American cotton through three deliveries.

The object of all the processes thus far described has been that of cleaning (in the picker), arranging the fibers in a parallel position to each other, making uniform, and drawing out the stock. In every case the stock delivered from a machine is lighter than when fed into it, and contains just twist enough to hold it together and prevent its being stretched or strained when unwound from the bobbin, and fed into the next machine. The minimum amount of twist in roving is desirable for the reason that it permits the stock to be drawn out more easily and uniformly, the little twist that is put in the roving by the slubber being practically eliminated when it is passed through the rolls of the intermediate. The same applies in the case of the roving passing from the roving to the spinning frame.

Fly Frames

The process in the manufacture of yarn after the cotton has passed through the drawing frame consists of further attenuation of the sliver, but as the cotton sliver has been drawn out as much as is possible without breakage, a small amount of "twist" is introduced to allow of the continued drawing out of the sliver.

From the drawing frame, the drawing passes through two, three, or four fly frames, according to the number of yarn to be made. All these machines are identical in principle and construction, and differ only in the size of some of the working parts. They are the slubber, intermediate, roving,—and fine or jack frame-fine, and the function of each is to draw and twist.

ROVING DEPARTMENT

Slubber machine, showing sliver of cotton passing through the rolls and then given a twist while it is wound on the bobbin.

Intermediate Frame

The function of the intermediate frame is to receive the slightly twisted rove from the slubber and add thereto a little more twist and draft. The rove is taken from two bobbins to one spindle in the machine, an arrangement which tends to insure strength and uniformity. The principle of the machine is in other respects the same as that of the slubbing frame.

Roving Frame

The function of the roving frame is to receive the twisted rove from the intermediate and add more twist and draft, thereby further attenuating the rove. As in the intermediate frame the rove is generally taken from two bobbins for one spindle.

Fine or Jack Frame

This machine is used when fine yarns have to be made. It is built on the same principle as the preceding frames, the only difference being that a finer rove is made from which finer numbers of yarn can be spun. As in the slubber, intermediate, and roving frames, the rove is taken from two bobbins for one spindle.

Spinning

In the manufacture of single ply yarn the final process is that of spinning, which consists in drawing out the cotton roving to the required size, and giving it the proper amount of twist necessary to make the yarn of the required strength. While the spinning frame is built on entirely different principles from the roving, intermediate, or slubber frame, the object of each machine is the same as that of the spinning frame. The principal point of difference is the amount of twist imparted to the cotton roving.

THREAD AND COTTON FINISHING

Thread

In general a twisted strand of cotton, flax, wool, silk, etc., spun out to considerable length, is called thread. In a specific sense, thread is a compound cord consisting of two or more yarns firmly united by twisting. Thread is used in some kinds of weaving, but its principal use is for sewing, for which purpose it is composed of either silk, cotton, or flax. Thread made of silk is technically known as sewing silk; that made of flax is known as linen thread; while cotton thread intended for sewing is commonly called spool cotton. These distinctions, while generally observed by trade, are not always maintained by the public.

The spool cotton of today is of a different grade from that made before the sewing machine came into general use. The early thread was but three cord, and contained such a large number of knots, thin places, etc., that it could not be worked satisfactorily on the machines, so manufacturers were called upon to produce a thread that would be of the same thickness in every twist. This was effected by making the thread of six cords instead of three, thereby producing a smoother and more uniform strand.

Manufacturing Processes

The raw cotton for the manufacture of thread must be of long staple. If the fiber is short the thread made of it will be weak, and hence unsuited for the purposes required of it. Ordinary cotton is not adapted to the manufacture of the better grades of spool cotton on account of the shortness of its fiber. Egyptian and Sea Island cotton are used because they have a much longer fiber and are softer in texture. The raw cotton comes to the factory packed in great bales, and is usually stored away for some months before it is used. The first step in the conversion of the bale of cotton into thread consists in giving the fiber a thorough cleaning. This is accomplished by feeding it to a series of pickers which pull the matted locks and wads to shreds, beat out the dirt and seeds, and roll the cotton in the form of batting upon cylinders until it issues from the finisher lap machines as a downy roll or lap.

The lap of cotton then goes to the carding rooms, where it is combed into parallel fibers by means of a revolving cylinder covered with fine wire teeth, sometimes 90,000 of them to the square foot. On leaving the carding machines the lap has become a gossamer-like web thirty-nine inches broad. This web is next passed through a small "eye" which condenses it into a narrow band about an inch in width, known as the sliver. By this time the fiber has been so drawn out that one yard of the original lap has become 360 yards of the sliver. The sliver now looks almost perfect, but if it were spun it would not make good thread. It is necessary to lay every fiber as nearly parallel as possible, so that there will be an equal number of fibers in the strand per inch. Besides this, the remaining dirt and short fibers must be removed and the knots and kinks in the fibers straightened out. To accomplish these objects the cotton must be "combed." First, the slivers are passed through several sets of rollers, each set moving faster than the preceding, so that the strands are drawn out fine and thin. In this condition the cotton passes to a doubling frame, and from thence to the lapping frame, a device combining six laps into one and drawing the whole out into one fine, delicate, ropy lap.

FINISHING ROOM

Cotton Finishing. Cotton fabrics, like other textiles, after leaving the loom must be subjected to various finishing processes so as to bring them into commercial condition. On piece-dyed goods part of the finishing is done before and part after the dyeing process. Each class of fabrics has definite finishing processes. In some cases weighting materials are added to the fabric so as to hide more or less its actual construction. Cotton fabrics just from the loom present a soft and open structure, more so than other textiles. Therefore it is necessary to use proper finishing materials and processes which will fill up the openings or interstices as produced in the fabric by the interlacing of warp and filling, and at the same time give to the fabric a certain amount of stiffness. Of course this finish will disappear during wear or washing, it having been imparted to the fabric to bring the latter into a salable condition. Cotton fabrics after weaving may be subjected to the following sub-processes of finishing: Inspecting, Burling and Trimming, Bleaching, Washing, Scutching, Drying. After the cloth leaves the loom it is brushed; then it passes over to the inspection table in an upward receding direction, so that the eye of the operator can readily detect imperfections. The ends of two or more pieces as coming from the loom are sewed into a string for convenient handling in the bleaching.

Bleaching

The object of bleaching is to free the cotton from its natural colour. The ancient method of bleaching by exposure to the action of the sun's rays and frequent wetting has been superseded by a more complicated process involving the use of various chemicals. Pieces of cloth are tacked together (sewed) to form one continuous piece of from three to one thousand yards in length. The cloth is next passed over hot cylinders or a row of small gas jets to remove all the fine, loose down from the surface. The goods are then washed and allowed to remain in a wet condition for a few hours, after which they are passed through milk of lime under heavy pressure, followed by rinsing in clear water. The goods are next "scoured" in water acidulated with hydrochloric acid, and boiled in a solution of soda, then washed as before in clear water. Next they are chlorined by being laid in a stone cistern containing a solution of chloride of lime and allowed to remain a few hours. This operation requires great care in the preparation of the chloride of lime, for if the smallest particle of undissolved bleaching powder is allowed to come in contact with and remain upon the cloth it is liable to produce holes. The goods are then boiled for four or five hours in a solution of carbonate of soda, after which they are washed. They are again chlorined as before and washed. The long strips are finally scoured in hydrochloric acid, washed, and well squeezed between metal rollers covered with cloth. After squeezing and drying, the cloth, if required for printing, needs no further operation, but if intended to be marketed in a white state, it must be finished, that is, starched or calendered.

Starching

The starch is applied to the cloth by means of rollers which dip into a vat containing the solution, while other rollers remove the excess. Sometimes the cloth is artificially weighted with fine clay or gypsum, the object being to render the cloth solid in appearance.

Calendering

The cloth is now put through the calendering machine, the object of which is to give a perfectly smooth and even surface, and sometimes a superficial glaze; the common domestic smoothing iron may be regarded as a form of a calendering utensil. The cloth is first passed between the cylinders of a machine two, three, or four times, according to the finish desired. The calender finishes may be classed as dull, luster, glazed, watered or moire, and embossed. The calender always flattens and imparts a luster to the cloth passed through it. With considerable pressure between smooth rollers a soft, silky luster is given by equal flattening of all the threads. By passing two folds of the cloth at the same time between the rollers the threads of one make an impression upon the other, and give a wiry appearance. The iron rollers are sometimes made hollow for the purpose of admitting steam or gas in order to give a glaze finish. Embossing is produced by passing the cloth under heated metal rollers upon which are engraved suitable patterns, the effect of which is the reproduction of the pattern upon the surface of the cloth.

Mercerizing

This is a process of treating cotton yarn or fabrics with caustic soda and sulphuric acid whereby they are made stronger and heavier, and given a silky luster and feel. The luster produced upon cotton is due to two causes, the change in the structure of the fiber, and the removing of the outer skin of the fiber. The swelling of the fiber makes it rounder, so that the rays of light as they fall upon the surface are reflected instead of being absorbed. The quality and degree of luster of mercerized cotton fabrics depends largely upon the grade of cotton used. The long-staple Egyptian and Sea Island cotton, so twisted as to leave the fibers as nearly loose and parallel as possible, show the best results. If the yarn is singed the result is a further improvement. Yarns and fabrics constructed of the ordinary grades of cotton cannot be mercerized to advantage. The cost of producing high-grade mercerized yarn is about three times that of an unmercerized yarn of the same count, spun from the commoner qualities of cotton. Mercerized yarn is employed in almost every conceivable manner, not only in the manufacture of half-silk and half-wool fabrics, and in lustrous all-cotton tissues, but also in the production of figures and stripes of cotton goods having non-lustrous grounds. Mercerized yarn used in connection with silk is difficult to detect except by an expert eye.

Characteristics of a good piece of Cotton Cloth. A perfect cotton fiber has little convolutions in it which give the strong twist and spring to a good thread. In this respect the Sea Island cotton is the best. There are five things requisite for cotton cloth to be good, *viz.*:

1. The cloth must be made of good fiber, that is ripe and long.
2. The fiber must be carefully prepared. All the processes must be well performed—for the very fine thread fiber must be combed to remove poor fiber. The combing, however, is not always done.
3. The warp and woof threads must be in good proportion.
4. The cloth must be soft, so that it will not crease easily.
5. It must be carefully bleached—the chemicals used must not be strong.

COTTON FABRICS

Albatross

Cotton albatross cloth is a fabric made in imitation of a worsted fabric of the same name. It has a fleecy surface. The name is taken from the bird whose downy breast the finish of the fabric resembles. The warp is usually 28s cotton, the filling 36s cotton. It is a plain weave. Filling and warp count 48 picks per inch. The goods are finished by being burled, sheared, washed, singed, dyed, rinsed, dried, and pressed, care being taken not to press too hard. Sometimes singeing is omitted. Albatross cloth is generally in white, black, or solid colours. It is not often printed. It is light in weight, and is used for dress goods.

Awning

A cotton cloth used as a cover to shelter from sun rays.

Batiste

Batiste is of French origin, and is a light, transparent cloth, made from a fine quality of combed cotton yarn. There is a gradual variation in quality ranging from a comparatively coarse to a very fine fabric. The variety of qualities will suggest some idea of the utility of the fabric. Its uses are even more varied than are the qualities. The finer grades are used for dress goods and all kinds of lingerie for summer wear, etc., while the cheaper grades are used for linings in washable and unwashable shirt waists. Batiste is woven in the gray, that is, with yarn direct from the spinning frame, with the exception that the warp yarn is well sized, in order to stand better the strain to which it is subjected during the weaving process.

Bourrette

A light weight, single cloth fabric, with two-ply cotton warp and wool or a combination of cotton and shoddy filling, made with the plain weave and in appearance a semi-rough-faced woollen fabric with fancy effects in twist scattered about it. It is used principally for ladies' fall suitings.

Bedford Cord

This is one of the most popular types of fabrics, the distinguishing effect being a line or cord running lengthwise of the cloth, the cord being more or less prominent. The cloth is made of cotton, or sometimes of worsted. The face effect of the Bedford cord is generally plain. Occasionally twill-faced cords are used. The cords vary in width from about one twentieth to one quarter of an inch. To get extra weight without altering the appearance of the face, extra warp yarns, termed wadding ends, are inserted between the face weave and the filling, floating at the back of the rib. When these wadding ends are coarse, they give a pronounced rounded appearance to the cord. They run from 88 to 156 picks to an inch.

Buckram

Buckram is derived from Bokhara. It may be described as a coarse, glue-sized fabric, and is made of cotton, hemp, linen, or cotton and hair (coarse) yarns, usually from 10s to 25s. Made of a double cloth warp, 22s cotton, 34 picks to the inch, for the face or top fabric 1/12's; weight from loom 2.22 ozs. per yard.

Bottom fabric 1/12's cotton; filling 1/16's cotton; 12 picks to the inch. Weight per yard, 1.8 ounces. These fabrics depend a great deal on the finishing. The men's wear requires less sizing on account of the hair it contains. The goods are piece dyed. Buckram is used principally for stiffening garments, and to give them shape or form. It is placed between the lining and the surface cloth of the garment in particular parts, such as the lapels, etc. It is used in the millinery trade, and is made into hats. Millinery buckram is sized two or three times. Calico takes its name from Calicut, a city in India, where cloth was first printed.

The majority of inexpensive cotton fabrics are constructed on the one up, one down system, or plain weave. Calico is no exception to this rule. The printed designs on calicoes may be somewhat elaborate or they may be simple geometrical figures.

In order, however, to comply with the true principles of art, such fabrics as calicoes should have but simple geometrical figures for their ornamental features. New styles and combinations of colours are produced every month and faster and lighter colour printed each season.

Most of the designs for calicoes and cotton cloth printing are made in Paris. At present the steam styles are most prominent; they are the fastest and lightest to be obtained. Calico is a printed cloth, the printing being done by a printing machine which has a rotating impression cylinder on which the design has been stamped or cut out.

The cloth in passing through the machine comes in contact with the impression cylinder. The cylinder revolving in a colour trough takes up the colour and leaves the impression of the design on the cloth. Calicoes may be seen in almost any colour. The printing machine is capable of printing several colours in one design. Calicoes, however, are usually in two colours, that is, one colour for ground and the other for figure. The ground colour in most cases is effected by dyeing the cloth in some solid colour. After the cloth is dyed the design is printed on it. The cloth, after it comes from the loom, is singed and bleached, then sheared and brushed to take away all the lint, and then sent to the dye house.

The first process there is to boil it, after which it is immersed in the dye tub. Calicoes are usually given what may be termed a "cheap cotton dye." By "cheap cotton dye" is meant that the colours are not fast, but will run or fade when subjected to water.

After the fabric is dyed, it is given to the printer, who ornaments the face of the cloth with some geometrical design; then it is practically ready for the merchant. After printing, the cloth is dried and steamed to fix the colour, afterwards soaped, washed, finished, and folded. The printing machine turns out about 400 to 800 fifty-yard pieces a day. Calico is used for inexpensive dresses, shirtwaists, wrappers, etc.

Cambric

Cambric is a heavy, glazed cotton fabric with a smooth finish. It was first made in Cambrai, France. It has a plain weave and a width of thirty-six inches. Cambrics are dyed in a jig machine. After dyeing they are run through a mangle containing the sizing substance, then dried, dampened, and run through a calender machine. The glossy effect is obtained in this last finishing process. Cambric is used for shirtwaists, dress goods, etc. The finer grades are made from hard twisted cotton of good quality.

Canvas

This is a term applied to heavy, plain weave cloths made with ply cotton yarn. They are used for mail bags, covering for boats, etc.

Chambray

Chambray is a staple fabric of many years standing, being next in rank among cotton goods after the better grade of gingham. Chambray is a light-weight single cloth fabric that is always woven with a plain weave, and always has a white selvedge. In effect it is a cloth having but one colour in the warp, and woven with a white filling, this combination producing a solid colour effect, the white filling reducing any harshness of warp colour in the cloth. It is composed of one warp and one filling, either all cotton, cotton and silk, or all silk.

It is twenty-seven to thirty inches in width and single 30s cotton warp to single 60s silk, the count of yarn being governed by the weight per yard desired. The weight per finished yard is two to three and one-half ounces. Good colours for the warp are navy blue, dark brown, lavender, black, nile green, etc.

When made of cotton warp and filling the fabric receives a regular gingham finish.

The loom width can be restored by tentering or running the goods over a machine fitted underneath with a series of coils of steam pipe. The top of this machine is fitted with an endless chain with a row of steel needles standing erect upon its face. Chains are adjusted to the width desired, and as the machine runs, both selvedges are caught by the needles and the cloth stretched to the required width.

Cheese Cloth

This is a thin cotton fabric of light weight and low counts of yarn, which ranks among the cheapest in cotton goods. It is used for innumerable purposes. The bleached fabric is used for wrapping cheese and butter after they are pressed. It is also much in demand for bunting for festival occasions, light curtains, masquerade dresses, etc. When used for bunting, draperies, and the like it is usually in colours, red, blue, cream, and yellow seeming to have the greatest demand. The weave is one and one or plain weave.

Chiné. Sometimes applied to glacé silk, or cotton two-toned effects. The name is French, meaning woven so as to have a mottled effect.

Chintz. Printed cotton cloth, with large, many-coloured designs, used for furniture covering. The Hindoo wears it as a body covering. Chintz is the Hindoo word meaning variegated.

Cotton Flannel. Napped cotton flannel. Made first for trade in Canton, China.

Crash. A plain fabric for outing suits, towels, etc.

Crêpe. A fine, thin fabric of open texture made of cotton.

Crepon. Large designs in figured crêpe. The name applies to the crispiness of the finish and is from the French word crêper, to make crisp.

Cretonne. Heavy cotton cloth printed in large designs, for drapery and furniture use. Cretonne was a Frenchman who first made the cloth.

Crinoline

Crinoline is a fabric composed of cotton warp, horsehair filling, or all cotton yarns. It is sold in varying widths, and is used by tailors and dress-makers in stiffening clothing. It is a cheap cloth of low texture and simple construction, the distinguishing feature being the stiff finish with either a dull or highly glazed face on the cloth.

Damask

A cloth of silk and cotton, silk and linen, silk and wool, or all linen in flowered or geometrical designs for drapery or table covering. The weaves used are mostly twills and sateens. It takes its name from Damascus, where it was first made.

Denim. This is a strong fabric usually made with a two up and one down twill. It is used for overalls, furniture covering, and floor covering.

Diaper. A figured cotton or linen fabric, which gets its name from the Greek diapron, meaning figured. It is generally of good quality as it is subject to excessive washing.

Dimity

A light-weight cotton fabric, the distinguishing feature of which is the cords or ribs running warpwise through the cloth, and produced by doubling the warp threads in either heddle or reed in sufficient quantity to form the rib desired. The name is from a Greek word meaning two-threaded. Dimity is a ladies' summer dress fabric, and is made of regular cotton yarn, from 1/60's to the finest counts in both warp and filling. It is made in both white and colours, solid white being used in the most expensive grades. Colours are often printed upon the face of the fabric after it has been woven in the white.

Domet

This cloth is napped similar to a cotton flannel. It is used for shirts, pajamas, etc., and made with bright coloured stripes and check patterns. The name is from domestic, home made.

Duck

Duck is a heavy single cloth fabric made of coarse two-ply yarn and of a plain weave. It derives its name from its resemblance to a duck's skin. It is of a lighter weight than canvas. In finishing duck is taken from the loom and washed and sized, then dried and pressed. If a fancy solid colour is desired the goods are dyed in the piece after the first washing. Duck is used in the manufacture of sails, tents, car curtains, and for any purpose requiring a good water-tight fabric, which will withstand rough usage. Duck has a stiff hard feel, and excellent wearing qualities. The lighter weights are used for ladies' shirtwaist suits, men's white trousers, etc.

Drill. A cotton fabric of medium weight generally made with the two up and one down twill. It is extensively used for shoe linings.

Eolienne is the name applied to a fine dress fabric characterized by having the filling of a much coarser count than the warp, thus producing a corded effect across the breadth of the goods. This class of goods is made up of a raw silk warp and either cotton or worsted filling, with the warp ends per inch greatly in excess of picks per inch. The goods are made up in gray, then dyed in the piece in any colour the trade desires. The darker shades find most favor for fall and winter use, while the lighter shades are preferred for summer wear. The width is from twenty-seven to fifty inches, and the price per yard varies from 85 cents to $1.25.

Etamine

An etamine is a thin, glossy fabric used principally for women's dress goods. Being a common and popular material for summer wear, it is usually made as a piece-dyed fabric. A good reason for making it piece-dyed is that this method is much cheaper than if the yarn is dyed previous to the weaving. Etamines were originally made with worsted yarns, which of course are more expensive; however, if a good quality of cotton is used, there is little difference in appearance between worsted and cotton etamine. The difference is chiefly in the wearing quality, worsted being more durable. The principal characteristic of an etamine is a crisp, glossy, and open structure.

Flannelette is a narrow, light-weight fabric composed of all cotton yarn, the filling being soft spun to permit of the raising of a very slight nap on the back of the goods. The cloth is woven with bleached yarn (warp and filling), the colour effects being afterwards printed upon the face of the goods by the printing machine. Flannelette is made with simple one or two coloured stripe patterns, either black and white or indigo blue and white, and in imitation of a Jacquard pattern. The finished fabrics are sold by the retailer at from eight cents to twelve and one-half cents per yard, are twenty-seven inches wide, and are used very extensively in the manufacture of ladies' wrappers, kimonos, etc., for house wear.

Fustian

A corded fabric made on the order of corduroy and used in England for trouserings, etc. First made at Fustat, a town on the Nile, near Cairo. Velveteen and cordings in the lower, coarser grades were sometimes called Fustian.

Galatea Cloth

Galatea cloth has been somewhat in demand in recent years by women requiring serviceable and neat-appearing cotton fabrics at a medium price. It is usually finished twenty-seven inches wide and retails at fourteen cents to twenty cents per yard. It is shown in plain colours as well as in figures, and in dotted and striped designs on white and coloured grounds. The patterns are obtained by printing. Some manufacturers have found that they can take a standard type of fabric and extend its use by varying the process of finishing. The base of the cloth—that is, the fabric previous to dyeing or printing or bleaching—is nothing more than an ordinary 5-end warp sateen of fair quality.

Gauze. A veiling net, made in Gaza in Palestine.

Gingham. Gingham is a single cloth composed entirely of cotton, and always woven with a plain weave. It is yarn-dyed in stripes or checks and was originally of Indian make.

It is the most widely known fabric on the market and is made in various grades, having from fifty to seventy-six ends per inch in the reed, and of 1/26's to 1/40's cotton yarns in both warp and filling. It is a wash fabric, made in both check and plaid patterns into which an almost unlimited variety of colour combinations are introduced. Ginghams are made with from two colours, warp and filling, to eight colours in warp and six in filling. Ginghams are used most commonly in the manufacture of ladies' and children's summer dresses and aprons.

Italian Cloth is a light, glossy fabric made from cotton and worsted, cotton and wool, cotton and mohair, and all cotton. It is used for linings for the heavier styles of ladies' dresses, also for underskirts, fancy pillow backs, etc. The cloth is woven in the gray undyed yarns. In the finer grades the warp is sized so as to facilitate the weaving process.

Jaconet. A thin cotton fabric, heavier than cambric. If properly made one side is glazed. Derived from the French word jaconas.

Khaki. Twilled cotton cloth of a brown dust colour, first used for men's clothing in India. The word khaki is Indian for earth, or dust-coloured.

WOOLLEN SPINNING

Woollen spinning consists of three main steps: drafting, twist insertion, and winding onto the cop. There are two major types of woollen spinning machine: mule spinning and ring spinning.

The main difference between them is that these three steps are performed differently by each machine and on the mule they occur in a discontinuous way whereby the drafting and twist insertion occur separately from winding onto the cop.

The ring-spinning machine combines them into a continuous sequence and so is more productive.

Fig. Wool Spinning Machine

Drafting Against Twist

Drafting serves two purposes; firstly it straightens the fibres in the slubbing so that their extent along the yarn axis is increased and secondly it reduces the linear density of the slubbing. Straightening the fibres increases the yarn strength (slightly at the expense of extensibility) while reducing the linear density at spinning allows a heavier slubbing to be produced at carding effectively increasing the productivity of the process.

On both the mule and the ring-frame each slubbing is fed from the creel via a surface drum to a pair of nip rollers that feed it into the drafting zone. Within the drafting zone a low level of twist is inserted into the slubbing and it is stretched by being drawn to the next stage by either the relative motion of the spindle as in the mule or by another pair of nip rollers running faster than the feed rollers, as in the ring frame. Woollen draft ratios are typically 30% and almost never higher than 40%.

Drafting against twist controls the rate of drafting of each segment of yarn within the draft zone. Without twist there would be a tendency for thin places to draft more than thicker places, making them thinner still; causing more unevenness in the yarn and frequent slubbing and yarn breaks. Twist tends to run preferentially to the thinner parts of a slubbing because the thinner areas have lower torsional rigidity. The relatively higher twist increases the strength of the thinner areas and so provides a stabilising mechanism to counteract their relative weakness. In theory the thicker places could be preferentially drafted over the thin and as they become lower in linear density the twist redistributes to control further drafting, in this way it is possible for the drafted slubbing to have better evenness than the parent slubbing. In practice the yarn irregularities have been shown to be mostly due (about 80%) to the web irregularities already present at the carding stage and that conversion from web to slubbing and to yarn generally only makes things slightly worse. The twist level used during drafting and the draft level itself have a marked effect on the yarn quality.

The optimum draft is usually around 30 to 35 % and depends on fibre length, fibre orientation, and slubbing uniformity.The optimum twist level depends on fibre-fibre friction and fibre orientation and length. The twist level is often not always directly proportional to the twister speed relative to the delivery speed as the twister can slip against the slubbing once twist reaches a certain level and resist further insertion of twist. The mule spinner has much lower productivity than the ring frame due to its discontinuous action but many manufacturers believe that it produces a more even yarn and it is still popular in the woollen sector.

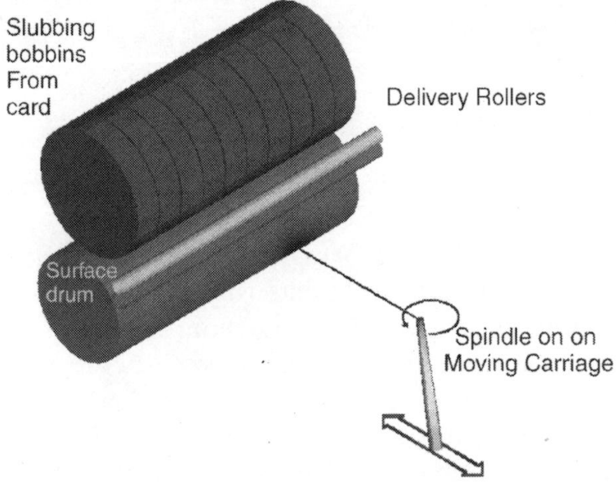

The mule frame is a horizontal machine and has a much longer draft zone in which the twist is inserted via the spindle. The spindle axis is almost perpendicular to the slubbing plane but is tilted forward slightly. During drafting and twist insertion as the spindle rotates the yarn flips over the top of the spindle inserting twist into the slubbing rather than winding it onto the cop. The flicking action is believed by some to help redistribute twist and set up vibrations that assist the drafting process.

In the first stage of the cycle slubbing is delivered from the creel via surface drums and the carriage moves back at the delivery speed while the spindles rotate to insert a small amount of twist. At a predetermined point in the carriage progress backwards the delivery rollers stop while the carriage continues so that the length of twisted slubbing starts to draft while twist continues to be inserted. At the end of the draw the spindle speed increases to its maximum to insert the final yarn twist and then the yarn is wound onto the package as the carriage returns to the start.

During winding a bar holds the slubbing down so that the spindle rotation causes the twisted yarn to be wound onto the cop. It also controls the yarn package build pattern. The longer draft zone on the mule may contribute to a better drafting action since the extra length allows a larger range of density fluctuations to exist within the draft zone simultaneously. Since the draft against twist mechanism is a competitive one, longer scale-length fluctuations can be handled in a longer draft zone. A schematic diagram of the woollen ring spinning system. The "False Twister" is so called because it inserts twist temporarily. It rotates as slubbing is drawn through it and so twist propagates up-stream. As each section of twisted slubbing passes through the point of twist insertion the sense of rotation of the twister relative to it reverses and each turn that was inserted up-stream is removed as it passes downstream of the twister. The slubbing is therefore only twisted up stream of the twister and the draft is only controlled in the upstream region. The twister is designed and positioned to minimise the untwisted length between it and the nip of the draft rollers, if this length is much less than the fibre length then little uncontrolled draft is can occur Some uncontrolled drafting will still happen.

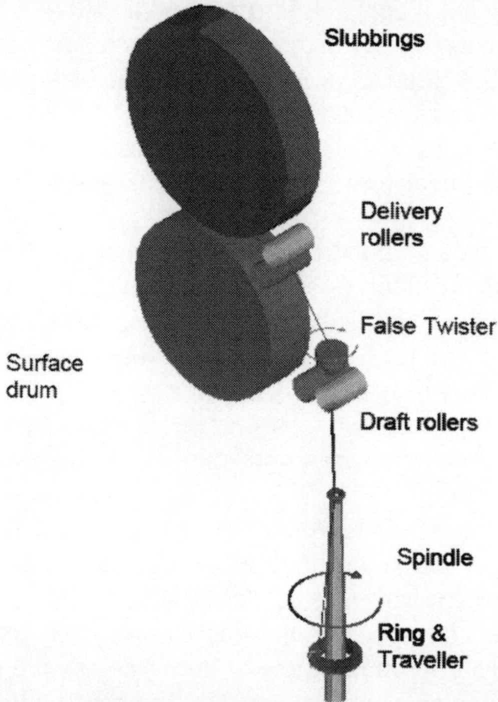

Fig. Schematic Diagram of Ring Spinning Frame

This is another difference between the mule and the ring frame and the most likely reason for better drafting behaviour on the mule; the mule-spun slubbing is twisted along its entire length during drafting. A system has been devised (and was commercialised by the Delpiano company of Italy) whereby the front draft rollers are housed in a cage which itself rotates so that the twist insertion point is nip of the front draft rollers and there is no region of zero twist. It has the added advantage that the drafting twist is completely controlled; a frictional false twister can slip and so the actual twist level inserted is unknown. This was shown to produce a more even slubbing than conventional drafting and was claimed to be at least as good as mule-frame drafting. The system was more complex than conventional frames and some difficulties could arise with end breaks and roller wrapping.

A wide range of false twisters have been used but they all seem to produce similar yarn qualities. Some twisters have knobs on their top surfaces designed to imitate the flicking behaviour of the mule. In these types the angle of the slubbing with respect to the twister axis is sometimes adjustable to vary the degree of flicking. It also changes the interaction between the twister and the slubbing and friction around it and so it is not known which effect is prominent when the angle is changed.

In ring-spinning the yarn twist insertion and the winding steps are combined. The ring has a lightweight traveller clipped to it through which the yarn passes on its way to the spindle cop. The traveller is dragged around the stationary ring by the rotating yarn and the friction between the traveller and ring means that the traveller turns slightly slower than the spindle so that yarn is wound onto the cop. As the spindle speed increases, the traveller-ring friction increases and centrifugal force on the yarn increases and so yarn tension increases.

The traveller weight must be chosen so that the tension it generates overcomes the centrifugal forces on the yarn at all speeds, otherwise it will fail to wind properly and a large "balloon" of yarn will form. If the traveller is too heavy then the tension may frequently exceed the local yarn strength at high speeds and the yarn will break at the thinnest places. Woollen ring-frame spindles usually have castellated tops and spiral loops positioned close to them so that the yarn balloon collapses onto the spindle before departing it again towards the traveller.

The smaller yarn balloon this generates compared to a free balloon reduces the spinning tension and allows higher speeds. The castellated spindle tops can generate extra yarn hairiness especially if spindle tops are damaged with knives when cleaned. Because the ring-frame is a continuous process it is much more productive than the mule-frame at the same spindle speed as the mule spindle is only spinning part of the time. However the mule frame is still quite popular with some traditionally minded spinners of fine woollen yarns. In worsted spinning and cotton spinning the mule has been completely superseded by the ring-frame and other newer systems such as rotor and air-jet.

The woollen yarn has a completely different and aesthetically pleasing character compared to worsted spun yarns. The fibres are less parallel and the yarns are hairier and so bulkier knitwear and flannels can be obtained from the woollen system. The woollen system can handle shorter fibres than the worsted system and is highly amenable to processing multiple fibre blends including large proportions of recycled fibres. The woollen card provides excellent intimate fibre blending but the speed is limited so that as the yarns go finer, the kilograms per hour produced drops and costs go up. Because of the disoriented conformation of the fibres in the woollen yarn a greater number of fibres in the cross-section is needed compared to worsted yarns. The spinning limit for woollen is about 90 fibres whereas for worsted it is 35 to 40 fibres. The finest yarns spun routinely on the woollen system are about 50tex, although 25tex woollen yarns are spun by some fine spinners from high quality fine wool inputs, eg 17 micron diameter and 60mm long lamb's wool.

YARN SPINNING

The invention relates to yarn spinning and more' particularly to the transmission of real twist from a spindle of a spinning system to a drafting zone to improve control of fibres during drafting. During the manufacture of spun textile yarns and the like the spinning process invariably includes a drafting stepr immediately prior to the insertion of twist into a yarn, to reduce the linear density of the feed material to that required in the finished yarn. In the yarn manufacturing system commonly known as woollen spinning the draft ^ ratios used are usually of the order of 1.3:1, and rarely exceed 1.5:1. Significantly higher draft ratios than this may affect the quality of the yarn adversely since the draft is limited by the high degree of interfibre entanglement which is an important feature of woollen slubbings (or slivers). This interfibre entanglement, and its effects, are somewhat variable along the length of the slubbing (or feed material) and this tends to cause irregular drafting, resulting in undesirable thick places and thin places in the yarn.

In order to minimise the effects of irregular drafting it is necessary to take steps to control the individual fibres during the drafting process in such a way as to draft the slubbing as evenly as possible at every point along its length. A number of methods are available to control the fibres but in the woollen spinning process the most commonly used method is to insert twist into the slubbing in the drafting zone. Twist travels along the yarn, or that section of yarn between the drafting rollers, until the torque is the same at all points within the drafting zone. The rate of travel of the twist is in excess of 38 m/sec"*[1], which is at least an order of magnitude greater than the throughput speed of the yarn. In order to balance the torque at all points in the drafting zone the twist migrates preferentially and rapidly to the thinnest place or places in the yarn, leaving the thicker places with relatively little twist. The average level of twist in the drafting zone is chosen such that the amount of twist which migrates into the thinnest places in the yarn is sufficient to bind the fibres tightly together, prevents fibres slipping relative to one another at that point and thus inhibiting further drafting at the thin places. The amount of twist remaining in the thicker places is insufficient to prevent fibre slippage, so these thicker places are caused to draft preferentially. Depending upon the quality and properties of the slubbing, the drafting process can then have the effect of improving the overall evenness of the yarn.

The twist used to control the fibres in the drafting zone can be real twist or false twist, and whilst real twist is generally preferred, it is far more common to use false twist because of the problems involved in passing real twist into the drafting zone.

Real twist is commonly used in the drafting zone of mule spinning machines, in which, by virtue of the design of both the machine and the process, it is simpler to use real twist than to use false twist, and this is one contributing factor to the generally superior quality of mule-spun yarns. Mule-spun yarns, however, are more expensive to produce than ring-spun yarns, so the latter are far more common in production.

In ring spinning, then, it is usual to use false twist in the drafting zone, and this twist is generated by a special mechanism included between a pair of front drafting rollers and a pair of back drafting rollers. The false- twist mechanism may be a false-twist spiral, a rotating tube or jaws or any rotating mechanism which grips the yarn is such a way that the yarn may be drawn through the mechanism by the front drafting rollers whilst the mechanism simultaneously inserts the desired level of twist upstream of the twisting point. All such mechanisms have attendant disadvantages including complexity in threading-up, the consumption of power etc., the inability to include false-twist right into the nip of the front drafting roller, and no facility to vary the average level of false twist to accommodate the longer term variations such as end-to-end variations in the slubbing.

It is also known to use real twist in the drafting zone of ring spinning machines with- ^ the aid of specially designed front drafting rollers which are individual to each spindle point and which are designed to rotate about their own axis in order to draft and deliver yarn to the spindle. but also to. rotate about the yarn axis such that real twist is transmitted from the balloon zone into the drafting zone. Such mechanisms, however, are generally expensive to manuf cture, complex to thread-up and absorb a considerable amount of power, they also do not include facility to vary the average twist level in the drafting zone according to the requirements of the individual ends of yarn.An object of the present invention is to provide means of transmitting real twist from a spindle area into a drafting zone of a spinning apparatus to provide fibre control between front and back drafting rollers during a drafting step.

Further objects and advantages of the invention are to ameliorate the above identified disadvantages and to simplify the threading-up operation of woollen ringframes; to reduce machine construction costs and operating costs; to provide means of varying the average twist level in the drafting zone according to the needs of individual ends of yarn; and to produce woollen spun yarns of generally improved properties.According to a first aspect of the invention there is-provided a method of transmitting real twist from a spindle area, through a nip of front drafting rollers, into a drafting zone, the method including the steps of: providing front drafting rollers, one of which is relieved around part of its circumference to provide an intermittent nip; dimensioning the relieved roller or rollers to control the amount of twist transmitted into the drafting zone; and drafting the yarn in the drafting zone.

According to a second aspect of the invention there is provided a means for drafting woollen slubbings, the means including a drafting zone between front drafting rollers and back drafting rollers, means for intermittently allowing a measured amount of real twist to run from a spindle area into the drafting zone, the arrangement being such that in use on a spinning machine the twist in the yarn within the drafting zone is intermittently topped-up to a preferred level. Further aspects of the invention which should be considered in all its novel aspects will become apparent from the following description which is given by way of example only.

Woollen slubbing 1 is drafted between back drafting rollers 2 and front drafting rollers 3, from whence it passes as spun yarn 7 via a lappet guide 4 and a ring and traveller mechanisms to be wound onto a rotating spindle. By virtue of rotation of spindle twist is inserted into the yarn at and this twist normally propagates upstream, against the flow of yarn, as far as the nip of the front drafting rollers.

One of the front drafting rollers is relieved around part of its circumference and near the middle of the roller space such that the slubbing will be intermittently nipped by rollers to effect an intermittent draft, and will also be intermittently released to allow twist to migrate from the spun yarn into the drafting zone.

Preferably the nip of the back drafting rollers the nip of the front drafting rollers and the lappet guide should be in-line to allow unimpeded propagation of twist from the spun yarn phase into the drafting slubbing phase when the nip between rollers opens.

The amount of twist propagating into the drafting zone depends upon the twist in the spun yarn and the time for-which the nip between rollers is open. When the nip between rollers is closed, twist will propagate with the yarn from the drafting zone back into the spun yarn phase.

The average level of twist in zone is chosen for optimum drafting according to the dimensions of drafting and the spinning system and the properties of the slubbing and its constituent fibres, and can be controlled by the twist level in the spun yarn the distance between the front drafting rollers and the pigtail, and the extent to which one of the rollers3 is relieved.

Referring now to front drafting roller (top or bottom roller) which includes a relief around part of its circumference and across the middle part of its width at which the yarn is normally nipped, leaving shoulders on either side of the relief to support the open nip and ensure the smooth running of the cooperating rollers. The relief is wide enough to accommodate the usual cross-traverse imposed upon the drafting slubbing for the purpose of spreading roller wear, and deep enough to accommodate the thickness of the yarn spun so that twist may propagate through the open nip unimpeded.

Fig. Yarn Spinning Machine

Referring now in cross-section-$_a$ preferred practical optional embodiment of the invention and which includes a second pair of front drafting rollers, one of which is also relieved around part of its circumference. In this embodiment of the invention the relieved rollers of roller pairs and preferably cooperate and are mechanically linked to ensure a fixed phase relationship is retained between them. The relieved rollers are thus preferably the bottom or driven rollers of each pair which are both relieved around less than half the circumference, and are driven at equal rotational speeds with the relieved portions operating precisely 180° out-of-phase. In operation then as the nip between rollers opens the nip between closes, and twist propagates from the yarn in the spindle area into the partially drafted slubbing in zone between the nips of rollers pairs. When the nip between rollers closes the nip between rollers opens, allowing the twist in zone to propagate onward into zone. Thus the average level of twist in the drafting zone can be precisely controlled by choosing the distance in zone between the nips of rollers.

A further advantage of this preferred embodiment of the invention is that by including two pairs of front drafting rollers the drafting action becomes continuous for when the nip between rollers is open, that between rollers is closed and drafting takes places between rollers and rollers; similarly when the nip is closed, and drafting then takes place between rollers. A further advantage of the present invention arises from the way in which real twist is propagated into the drafting zone and how that twist is controlled, in part, by the linear density of each individual end of slubbings. Thus in the case of a lighter-than average area in an end of slubbing a higher level of twist will be accepted into the drafting zone, tending to inhibit further drafting of that area.

A heavier-than-average area in the slubbing will accept a lower level of twist and this promotes easier drafting of such thick areas. Similarly constantly thin ends of slubbing can be expected to accept a higher average level of twist into the drafting zone resulting in some reduction in the actual average draft, whilst constantly thick ends of slubbing are able to accept less twist into the drafting zone and can be expected to draft more readily with a higher actual draft.By this invention is provided a method and apparatus by which real twist can be propagated from the spindle area into the slubbing in the drafting zone of a woollen drafting system* for the purpose of controlling the fibres during the drafting process.

Particular examples have been described for effecting the purpose of this invention which is to continuously transfer real twist from the spindle area into the drafting zone of a woollen drafting system, and it is understood that there can be variations in the methods used without departing from the spirit and scope of the invention. For example it is possible and practical to affect an intermittent nip, thereby allowing the propagation of twist upstream into the drafting zone, by using a standard, unrelieved roller nip but intermittently guiding the yarn into and out of that nip.

WORSTED SPINNING

PROCESS DESCRIPTION

Worsted spinning is the conversion of combed top into yarn.

The top sliver is progressively thinned by drafting the strand by a larger amount than the number of input slivers (doubling). The doublings help to reduce any irregularity present in the input material or introduced in the drafting process. When the sliver is thick the control of fibres is achieved using pins and as it thins better control is provided by balloon rollers and/or aprons gilling stages are used. The drawing stages are followed by one or two roving stages. A flyer rover consolidates the output strand by inserting a small amount of twist. A rubbing rover consolidates the strand by rubbing between reciprocating aprons.

The roving packages are mounted (creeled) on the spinning frame and drafted (by a factor of 14 to 30) before twist is inserted. For pure wool strands the drafting is achieved using 3 pairs of rollers with the middle pair having a central recess and driving a pair of aprons which lightly grip the strand. Twist is inserted in the drafted strand by having the forming yarn pass through and pull a metal loop (traveller) which slides around a lubricated ring. The whole package rotates within the balloon formed by the yarn. Each turn of twist in the yarn requires a complete rotation of the package (bobbin) and, for a fine weaving yarn, it is typical for 500 to 1000 turns/m to be inserted. The speed of rotation is limited by the tension that the forming yarn can withstand, with the tension coming from the friction of the traveller. Higher speeds require smaller rings and lighter travellers but the tension must also be enough to control the balloon.

Fig. Worsted Spinning Machine

Full yarn packages are removed (doffed) and fed to a winder either directly or after a steaming operation to reduce the twist-liveliness of the yarn. During the winding operation the yarn from successive small bobbins is joined to form a large package and thick, thin and coloured places can be removed (cleared). The clearer sensors cut the yarn when a fault is detected and then the two ends are located, overlayed and rejoined with a splice (or knot).

In order to produce a yarn that can survive weaving it is normal to twist two singles yarns about each other in the opposite direction (or the same direction if a stiffer crepe yarn is desired). Two or more yarns can be assembly wound onto the one package in preparation for this twisting operation or several packages can be directly mounted on the twister. For knitting a number of yarns may be wound and twisted together to produce a bulkier or torque-balanced yarn. The older method was to use another ring/traveller machine (ring twisting) but it is now more common to use a 2-for-1 twister in which the component yarns rotate around the suspended package(s) and 2 turns of twist result for each rotation. The final yarn may be steamed or conditioned before dispatch to the weaver or knitter.

Fig. Worsted Spinning Ring frame

RECENT DEVELOPMENTS

It has become almost standard for the spinning mill to re-comb tops of fine (<21ìm) wool irrespective of whether the top is dyed. Although expensive the improved spinning performance is claimed to outweigh the cost. Re-combing is also much better than gilling for achieving a uniform blend of different colours or with synthetics. On modern drawing machinery only 3 or 4 gillings of the top are required and any extra doubling and drafting stages (other than needed for blending) are better carried out when the sliver is thin and fibre control better, that is, at the roving rather than early stages.

It is possible to produce a singles yarn that is directly weavable without two-folding by spinning two separated strands (Sirospun) or by splitting the one strand into multiple components (Solospun). The latter allows slightly finer yarns than two-folding but the main advantages are in cost savings from removing a twisting step. The resultant yarns are leaner than two-fold yarns and still have uni-directional twist but the surface fibres are more strongly bound than for a singles yarn.

Yarns of lower hairiness and slightly higher tenacity can be produced using compact or condensed spinning. Here the strand is compacted with air in an additional zone after drafting but before twist is inserted. This gives improved spinning performance but the yarn is not sufficiently abrasion-resistant for use as a warp yarn in weaving.

Fig. Worsted Yarn Spinning

Collapsed balloon spinning is also available in which the yarn balloon is caught by a crown or bent finger on top of the spindle and winds partly around the bobbin. This greatly reduces the tension on the forming yarn above the spindle although the tension is higher at the traveller. Higher spinning speeds are possible but the empty bobbin tubes need to be of larger diameter and doffing is more complex.

Smaller diameter rings are used for finer yarns and diameters have reduced over time because smaller rings allow higher speeds but less yarn fits on the bobbin and so more doffing, piecening and splicing are required.

Automation has steadily increased with automatic doffing now routine. Ends-down detection is available as well as roving stop motions and spinning speeds can be adjusted according to the level of end-breakage. The bobbins can be automatically conveyed to a linked steamer and/or winder and such systems can monitor and report on which spindles are producing faulty yarn using clearer sensors which can measure evenness and hairiness. Splicing has generally replaced knotting on the winder with hot air splicing (Thermosplicer) being particularly effective.

Synthetic filaments such as polyester and acrylic can be blended after being cut or broken to a similar length to the wool although they may be dyed separately before blending. Bi-component blends can be made on the spinning frame using Sirospun and attachments are also available to feed a continuous filament, such as an elastomer to impart stretch, behind the front roller.

Although slow, ring spinning is dominant for wool yarns. Self-twist spinning has a small presence for hosiery yarns. Open-end or rotor spinning is used for some wool/cotton blends but the fibre must be shortened to match the size of the cotton rotors and fibre handling systems. Air-jet, including Vortex, spinning has been tried but is not in widespread use. Friction or DREF spinning is used for some coarse multi-fibre blends. In general, these competing systems cannot produce yarns which are as strong or with as few fibres in the yarn cross-section.

Processing Problems

The drawing of the top sliver down to roving should be straightforward using standard loadings and pinnings for the diameter of the wool and making small adjustments according to the mean fibre length. The quality at each stage can be checked in terms of the Index of Irregularity. Lapping of rollers can arise if the wool is too dry or if there are wet patches, or there is insufficient lubricant/antistat, or if scouring left excessive residuals, or if the rollers have become dirty or degraded.

The generic problem is one of suspected inferior yarn quality and excessive ends-down which have to be compensated for by slowing the spinning speed and, possibly, increasing the twist. The first thing to realise is that yarn properties and spinning performance are quite sensitive to small changes in fibre properties but expected performance can be predicted. However, if performance is below expectations then it can be due to a myriad of causes, for example, machine maintenance (worn rollers, rings, pig-tails or aprons, eccentric spindles, slipping drives, misaligned components, missing ring lubrication etc.) or machine settings (wrong apron spacer, wrong back-draft etc.) or faulty input material (fibre damaged in dyeing, irregular roving, slubs, bad joining etc.) or low moisture content (due to low RH%). Establishing the actual cause of underperformance can take considerable time but firstly requires the ability to measure yarn and fibre properties under standard conditions.

Problems can also become apparent in downstream processing but most should be picked up early if sliver and yarn properties are monitored. Excessive faults in winding can arise from as far back as faulty output webs in combing. Even at winding and two-folding care must be taken because faulty splice settings, or bad knots in two-fold yarn, may show up in weaving.

Reducing yarn hairiness with a modified yarn path in worsted ring spinning

The spinning geometry of a ring frame plays an important role, and the twist triangle is the critical region in ring spinning. Changes in the spinning geometry may affect yam properties. This paper examines the idea of ring spinning with a "diagonal" yam path, and the effect of such a path on yam properties, particularly hairiness. Both "left diagonal" and "right diagonal" yam arrangements are tried on a 24-spindle Cognetex FLC worsted ring frame. The hairiness results obtained from the Zweigle hairiness meter show that the right diagonal yam path produces yarns of lower hairiness than the conventional ring spun yam in almost all the hair length groups. Yam evenness and tenacity are not as sensitive to the change in yam path. The mean spindle speed at break is also tested, and there is some deterioration in spinning efficiency with the right diagonal yam path, particularly at higher spinning speeds. Results from this study may help explain variations in yams spun on poorly aligned ring frames.

In manufacturing wool yams on widely used ring spinning systems, a considerable number of wool fibers are only partially bound into the yam structure. Many fiber ends and loops protrude from the body of the yarn, making it hairy. Yarn hairiness has a significant effect on yam strength, spinning efficiency, yam weaving, and knitting performance, as well as the properties of the resultant fabrics made from such yams, particularly their pilling propensity and uneven dye uptake. Nevertheless, ring spun yams are still regarded by most yam technologists as having the ideal blend of properties [5], and only ring spun yarns can be considered for finer yarn counts. Despite the apparent advantages of ring spun yams, there is still considerable room for improvement. If a ring-spun yam is examined under the microscope, it is easy to see that many fibers remain uncontrolled and are not bound into the yam structure. These protruding hair fibers make no contribution to yam strength and elongation. They are also often unwelcome in downstream processes and amount to deficient utilization of raw material. The volume of fly in spinning mills and in further processing is also mainly due to yarn hairiness. It is thus obvious that even ring spun yams are not yet ideal in terms of structure and quality. The quest for higher quality ring spun yams therefore continues, particularly in the area of hairiness reduction.

In recent years, major European textile machinery manufacturers, such as Rieter and Zinser, have invested heavily in the development of compact ring spinning systems that produce yams with reduced hairiness. These systems reduce the spinning triangle or twist triangle to a minimum to ensure improved fiber incorporation into the yarn structure and hence reduced yam hairiness. This highlights the importance of the spinning geometry in ring spinning. The term "spinning geometry" includes all distances, inclinations, and angles in the fiber flow between entering the draft system and yam take-up on the cop. Klein stated that the spinning triangle and spinning angle affect fiber breakage considerably, while the spinning triangle also influences yam structure. The effect of key geometric elements on spinning ends-down and yam quality.

It is obvious from these figures that the spinning triangle is the most important factor influencing yam quality and ends-down. For theoretical purposes, the triangle is often assumed to be symmetrical, but in reality this is not the case. For a Z-twist yarn, fibers on the right hand (RH) side of the triangle often undergo a pretwisting process that effectively binds the fibers, while fibers on the left hand (LH) side are less controlled. Often these fibers on the LH side of the triangle become hair fibers when the yam is formed. One hypothesis is that if we can shorten the distance the fibers on the LH side of the triangle have to travel before reaching the convergence point, yarn hairiness is likely to be reduced. There are two ways to reduce this distance. One way is to increase the twist level so that a smaller triangle is formed during spinning. Many researchers have verified that a yarn with a higher twist level is less hairy. Another approach is to modify the yam path. We have used that approach here with "left-diagonal" and "right-diagonal" configurations. Our experimental work in this paper evaluates the performance of these diagonal yam path arrangements in terms of yam quality and spinning ends-down.

THE SET-UP

The two possible ways of spinning with a diagonal yarn path-left diagonal and right diagonal. With the left diagonal arrangement, the yam issuing from a drafting unit is taken up by an adjacent bobbin to the left of the drafting unit, instead of the bobbin directly below it. For the right diagonal configuration, the yam from a drafting unit is taken up by an adjacent bobbin to the right of the drafting unit, instead of the bobbin directly below it. Reducing yam hairiness we postulated that for spinning a Z-twist yam, the left diagonal arrangement should give better control of the fibers on the left-hand side of the triangle (due to the reduced distance the less controlled fibers on the left-hand side have to travel to reach the convergence point) than with the conventional arrangement, while the fibers on the right-hand side of the triangle are still controlled by the pre-twisting process.

Therefore, with this left diagonal arrangement, yam hairiness should be reduced. On the other hand, the right diagonal arrangement may actually increase yam hairiness, because of further reduced control of fibers on the left-hand side of the twist triangle. We further stated that one drawback of the diagonal arrangement may be that the twist blockage at the pigtail guide will be more severe than the conventional set-up, which could lead to increased ends-down in spinning. But the spinning tension (with the diagonal yam path arrangements) above the pigtail guide may be lower due to the increased capstan effect between the yam and pigtail guide. These two factors may counteract each other, and the resultant effect on ends-down could be minimized. In this study, we further examine the various aspects of spinning with diagonal yarn path arrangements.

MATERIALS AND SETTINGS

We used a 24-spindle Cognetex FLC worsted ring spinning frame to process the pure wool yam. The parameters of fibers, roving, and yams are given in Table I. We used a draft of 23.27, 25# traveller, 7000 rpm spindle speed, and Z-direction twist in these experiments. Before commencing the spinning trial, we replaced all the travellers.

In order to evaluate the effect of spinning with a diagonal yam path on yam quality and spinning efficiency, we designed a series of experiments and tested the yams for hairiness (Zweigle 6565 hairiness meter), evenness (Uster tester

4), and strength (Uster Tenso— Rapid). The results of these experiments are reported next. Results and Discussion

Experiment 1

Experiment 1 was designed to compare the effect on yam hairiness of conventional ring spinning and spinning with the left-diagonal yam path. The Latin square experimental design was used to spin four lots of yarn in the order of conventional, left diagonal, left diagonal, conventional. Each lot consisted of twenty yarn bobbins. We conditioned all the samples for at least 48 hours under standard conditions (65 +/- 2% RH and 20+/-2 deg C) before testing the yarns. We selected the Zweigle 6565 hairiness meter for yam hairiness testing under standard conditions at a speed of 50 m/min and Sg pre-tension. We selected 100 rn as the testing length, tested five blocks of each bobbin, and calculated the average value for each bobbin. We tested ten bobbins from each lot. The hairiness results of four lots of yam are shown in Table II. Lots I and 4 represent conventional ring spinning, and lots 2 and 3 represent spinning with the left diagonal yam path. Long-length groups for yarns spun with the left diagonal arrangement are more than the hairiness readings of conventional ring spun yarns. These results are somewhat unexpected. With the left diagonal arrangement, the Z-twist yarn should have better controlled the fibers on the left-hand side of the triangle (due to the reduced distance) than the conventional arrangement, while the fibers on the right-hand side of the triangle are still controlled by the pre-twisting process. Therefore, with this left diagonal arrangement, yarn hairiness should be reduced. One possible explanation for the current results is that the two ring-spinning machines are different in the two experiments. The previous study involved a six-spindle spin tester, which has a different geometry than the Cognetex FLC worsted ring spinning frame in this experiment. Spinning with the left diagonal yam arrangement changes two spinning geometry factors-the spinning triangle and the spinning distance. The increased spinning distance and possibly the friction between the yarn and the yam pigtail guide might have contributed to the increased yarn hairiness with the left diagonal yam path. However, if these were the reasons for the increased yam hairiness with the left diagonal yam arrangement, one would expect a similar result for the right diagonal yam arrangement.

Bibliography

A.R. Israni and P.K. Shah: *Basic Mechanical Engineering*, B.S. Publications, Delhi, 2011.

A.R. Israni, S.M. Bhatt and P.K. Shah: *Elements of Mechanical Engineering,* BS Publication, Delhi, 2009.

Ajay Kumar Tamrakar and Dineshkumar Harursampath: *Wiley Acing the GATE Mechanical Engineering*, Wiley, 2014.

Anil Kumar and K. Sudhakar: *Fundamentals of Mechanical Engineering*, BS Publications, Delhi, 2012.

Basant Agrawal, C.M. Agrawal: *Basic Mechanical Engineering*, Wiley, Publication, Delhi, 2004.

Bhargava, A. K., Sharma C. P.: *Mechanical Behaviour and Testing of Materials*, PHI Learning, 2008.

Biomechanics - Mechanical Properties of Living Tissues, Fung, Springer, Publication, Delhi, 2007.

Crowder: *Electric Drives and ElectroMechanical Systems*, Elsevier, Publication, Delhi, 2009.

D R Subbu: *Dictionary of Mechanical Engineering*, SBS Publication, Delhi, 2009.

D R Subbu: *Encyclopedia of Mechanical Engineering*, SBS Publication, Delhi, 2007.

D.K. Singh: *Elements of Mechanical Engineering*, Ane Books, 2012.

D.K. Singh: *Multiple Choice Questions in Mechanical Engineering*, Ane Books, 2012.

Daniel E.: *Mechanical Assemblies : Their Design, Manufacture, and Role in Product Development*, Whitney, Oxford University Press, 2009.

Devendra Vashist: *Mechanical Engineering : Fundamentals* : , I.K. International, 2010.

Dhananjoy Shaw: *Mechanical Basis of Biomechanics*, Sports Publication, Delhi, 2007.

Donald M. Johnson, Joe Harper, David E. Lawver and Philip Buriak: *Mechanical Technology in Agriculture*, International Book Distributing Co, 2005.

Dukkipati And V. Rao Srinivas, J.:*Textbook Of Mechanical Vibrations*, PHI Learning, 2012.

E. Manouchehr Motamedi: *Micro Opto Electro Mechanical Systems*, New Age International, 2010.

G.H.F. Nayler: *Dictionary of Mechanical Engineering*, Jaico Publishing, Delhi, 2007.

H. Prashad: *Integral Approaches to Tribo-Testing in Mechanical Engineering*, BS Publication, Delhi, 2009.

Jack, Henry R. Busby and George H. Staab: *Mechanical Design of Machine Elements and Machines*, Wiley India, 2011.

John Bird: *Mechanical Engineering Principles*, Elsevier, Publication, Delhi, 2011.

K.C. John: *Mechanical Workshop Practice*, PHI Learning, 2008.

K.K. Ramalingam: *Handbook of Mechanical Engineering Terms*, New Age Publication, Delhi, 2009.

Lobontiu: *Mechanics of Microelectromechanical Systems*, Springer, Publication, Delhi, 2004.

Maitra: *Handbook of Mechanical Design*, Tata McGraw-Hill, 2004.

Manjula B Waldron and Kenneth J Waldron: *Mechanical Design : Theory and Methodology*, Springer, Publication, 2007.

Marc Andre Meyers and Krishan Kumar Chawla: *Mechanical Behavior of Materials*, Cambridge Univ Press, 2004.

Martin Fernandez: *Dictionary of Mechanical Engineering*, APH, Publiucation, Delhi, 2011.

P.N. Gupta and M. P. Poonia: *Elements of Mechanical Engineering*, Standard Publication, Delhi, 2009.

Pushpendra Sharma, Vishav Kamal and Prashant Sharma: Mukesh Pandey: *Mechanical Engineering Data Handbook*, I.K. Interrnational, 2015.

R.N. Iyengar: *Elements of Mechanical Vibration*, I.K. International, 2010.

Rasheed Ahmed Khan, Ali Hasan and Ravinder Kumar Anand: *Mechanical Engineering*, Atlantic, Publication, Delhi, 2009.

S. P. Venkateshan: *Mechanical Measurements*, Ane Books, 2015.

S.K. Sharma: *Judicial System in Mechanical Warfare*, Sumit Enterprises, 2012.

Trymbaka Murthy: *A Textbook of Elements of Mechanical Engineering* , I.K. International, 2006.

Uday Tusti: *Encyclopaedia of Mechanical Engineering*, Anmol, Publication, Delhi, 2006.

Waldron: *Mechanical Design: Theory and Methodology*, Springer, Publication, 2008.

William Andresen: *Die Cast Engineering: A Hydraulic, Thermal, and Mechanical Process*, Routledge, Publication, Delhi, 2010.

Index